THE WESTERN FRONTIER LIBRARY

(Complete list on page 211)

THE GREAT DIAMOND HOAX

THE
Great Diamond Hoax

and Other Stirring Incidents in the Life of

ASBURY HARPENDING

Edited by James H. Wilkins

with a Foreword by

GLEN DAWSON

Norman: University of Oklahoma Press

Library of Congress Catalog Card Number: 58–6853

Composed and printed at Norman, Oklahoma, U.S.A., by the University of Oklahoma Press.
First printing.

To my friend, unassuming John A. Finch, of Spokane, Washington, a man of great ability, possessing, according to my ideals, all the attributes of greatness, as a token of my deep esteem, this book is dedicated.

THE AUTHOR

FOREWORD

By Glen Dawson

THE "GREAT DIAMOND HOAX" was one of the most fantastic episodes in Western history. In this book Asbury Harpending claims to tell the "whole truth" about it, and in general his story has been accepted. However, the subject is still surrounded by mystery, inconsistency, and controversy. While there can be no question that Asbury Harpending participated in the events he describes with such pungency, there still remains the big question of the extent of his innocence or guilt in the Diamond Mine Swindle, called by Harpending the Great Diamond Hoax.

By far the greater part of Harpending's book is devoted to the author's "sixteen tempestuous years" from 1857 to 1873, the years from 1873 to 1913 being covered in one short chapter at the end. While the historian usually prefers an account written on the spot, yet Asbury Harpending in 1913 could recall and record the colorful incidents of his life, some of which he no doubt had told and retold orally before he put them down on paper. He could write with authority and a minimum of extraneous detail about events for which contemporary records are scant: the Confederate secret society known as the Knights of the Golden Circle, the *Chapman* case, Civil War days in Kern County, California, the financial deals with William G. Ralston, and the Great Diamond Hoax. The scene shifts from Kentucky to California, Mexico, the Confederate States, New York, London, Wyoming, and back to Kentucky.

The history of his book is interesting. In 1912 there were published two large volumes entitled *San Francisco,* by John P. Young, in which Asbury Harpending was termed a scamp, ingenious rogue, accomplished scoundrel, and the genius who conceived and elaborated the Diamond Swindle. It was to refute these charges, says Harpending, that *The Great Diamond Hoax* was written. The contents were first published in installments in the San Francisco *Bulletin* in daily issues from September 13 to October 28, 1913, the articles being arranged for publication by a special writer for the *Bulletin,* James Hepburn Wilkins. Wilkins, who died in 1934 at the ripe age of ninety, had been trained in engineering at the University of California but had switched to newspaper writing and editing, and in later life served on the state prison board and as mayor of San Rafael. The copyright on these articles was secured in the name of Asbury Harpending, who at that time was living in a house, which is no longer standing, at 2607 Fruitvale Avenue, Oakland. The first installment contains the following explanation:

> It is not my purpose to tell the story of my life, but only so much of it as touched on matters of historical importance and affairs of peculiar public interest. Most men slip through the world guided by a single purpose, so calmly that they scarcely make a ripple on the surface. But fate, together with a restless disposition and a love of venture, so ordered it that I should become an actor in the events of a moving epoch; events that not only shook California to its center, but reached across the continent to the Atlantic seaboard and beyond the ocean to the Old World. Some of the events have been mistold, half told, or not told at all. It is to throw a larger and truer light on history that I am furnishing my recollections. But in order that the narrative may be connected and not broken by halts for explanation, it is essential to begin a few

years ahead of, or rather behind, historic interest. It will soon be apparent that this is done not for personal glory, but to promote continuity and in the end to save the reader's time.

In 1915 the articles appeared in a 283-page book, including twenty-five portraits, bound in light-blue cloth, gilt stamped. Some of the copies had a dust jacket which proclaimed the price as $1.50. Although the James H. Barry Company is listed as publisher, without doubt Harpending himself paid the printing bill. There is no reference to the newspaper version, but the chapter headings are taken directly from the newspaper headlines and on the title page James H. Wilkins is named as editor. The book is dedicated to John Ayland Finch (born in England in 1854, came to Leadville, Colorado, in 1881, and to Spokane in 1887), whose firm of Finch and Campbell did much to develop the mining industry of the Pacific Northwest.

Asbury Harpending was unsuccessful as a publisher, and not long after the publication of *The Great Diamond Hoax,* he sold the remainder, some 1,200 copies, to Harold Holmes, well-known California bookseller, for five cents a copy. Mr. Holmes recalls stacking them in great piles in the doorways of his shop, priced at ten cents a copy, though he later raised the price to fifteen cents.

In a copy of *The Great Diamond Hoax* in the possession of the California Historical Society, the author has written a long inscription regretting that he did not dedicate the book to S. H. Daniels, whom he knew for over half a century:

> I will explain why so small a fraction of my most thrilling experiences appeared in the book. The book is a replica of the serials that appeared daily in the

> *Bulletin* for months, which paper, as you know, was interested only in California history, irrespective of the personality of the individual concerned. This serial could easily have been continued in the paper for three, or possibly four, months. Mr. Wilkins, who at that time was a member of the *Bulletin* staff, was instructed to limit the daily issue to 1,500 words to condense the whole story in such a way that only those incidents were given that were a direct bearing upon California history, leaving out most of the human touches, many thrilling incidents, in fact almost all of a personal nature. I have since prepared a history of my life in California and elsewhere, to publish some time. This book will be dedicated to S. H. Daniels, who was the real instigator. It is with great pleasure I present this book to Mr. S. H. Daniels with the compliments of the author.

Whether or not this "history" still exists in manuscript form, we do not know.

Asbury Harpending spent more of his life in California than the last chapter of his book would indicate. Finally, in 1918, he moved to New York, where he died on January 26, 1923, at the age of eighty-three. He had three daughters, one of whom married Alexander K. Pavellas, a Greek consul, and another of whom married George D. Papageorge Palladius, editor of a Greek newspaper.

Some years ago a lot of Asbury Harpending's old business papers were disposed of by a stamp dealer in San Francisco, and after passing through several hands, finally reached the California Historical Society. Many of the papers, unimportant in themselves, would reveal many details and dates omitted by Harpending. To date, this source has been only partially investigated, although the late Joseph Henry Jackson at the time of his death was contemplating a biography of Harpending.

Included among the papers is an elaborate telegraph code for use on the inspection trip to the diamond fields. Even more interesting is a business agreement between Philip Arnold and Asbury Harpending, dated October 31, 1871, which outlines the ownership of certain diamonds and various arrangements. This was six months prior to the date on which Harpending claimed he arrived in San Francisco for the purpose of investigating the diamond discovery. It is not my purpose in this foreword, however, to render a verdict on Asbury Harpending—indeed, at this writing not all the evidence has been presented. This is simply a reintroduction to Harpending's own story.

His account has been the basis of many articles and books, and will remain the prime source for whatever is to be written on the Great Diamond Hoax in the future. It would be difficult to estimate how many motion pictures, short stories, and novels have been inspired by this book. The subject has received attention from such writers as George D. Lyman, Herbert Asbury, Arthur Woodward, Robert O'Brien, Benjamin Franklin Gilbert, Miriam Allen de Ford, and Harry H. Crosby.

The Great Diamond Hoax was a favorite book of the late E. DeGolyer, of Dallas, Texas. This is sufficient recommendation for its inclusion as the tenth volume of the Western Frontier Library.

Los Angeles, California
January 15, 1958

Contents

Illustrations

THE GREAT DIAMOND HOAX

PREFACE

ON MY RETURN to California, after an absence of many years, my attention was called, for the first time, to the fact that my name had been associated unpleasantly with the great diamond fraud that startled the financial world nearly half a century ago. Plain duty to my family name and reputation compelled me to tell the whole story of that strange incident so far as my knowledge of it extends. I sincerely trust that a candid reading of these pages will satisfy the public that I was only a dupe, along with some of the most distinguished financiers of the last generation. Concerning two of the historians who maligned me, I am without redress. They are dead. The latest author, Mr. John P. Young, repeated the accusation of his predecessors in his history of San Francisco. This gentleman has admitted that he merely copied the story of the earlier works, having no personal knowledge of events at that period, and has handsomely admitted, over his signature, that he unconsciously did me an injustice.

To the diamond story I have added, at the request of friends, some of my experiences and reminiscences of California of the early days.

Asbury Harpending

CHAPTER I

Early Years—My Voyage to California

My father was one of the largest landed proprietors of Kentucky, in the southwestern section of the state. That was still on the frontier of the Far West. Beyond stretched the land of enchantment and adventure—the plains, the mountains, the unbroken solitudes, the wild Indians, the buffaloes, and the Golden State on the shore of the Pacific.

Youngsters whose minds are occupied today with baseball and tennis and who still retain a lingering love for taffy, sixty years ago on the frontier were dreaming of wild adventures that were nearly always realized to some extent. We lived on the border line, where the onward wave of emigration broke and scattered over the vast vacancies of the West, and it is hardly saying too much to assert that fully seven boys in ten were caught and carried forward with the flood before they had gone very far in their teens.

For myself, I simply gave up to the spirit of the times. At the age of fifteen I ran away from college to join an aggregation of young gentlemen but little older than myself, who enlisted under the banner of General Walker, the filibuster. The objective was the conquest of Nicaragua. The Walker expedition sailed to its destination, and what followed is a matter of well known history. But for my companions and myself, numbering 120 in all, it ended in a humiliating disaster. For, as we sailed down the Mississippi River, the long arm of Uncle Sam reached out and caught us, like a bunch of truant kids. I managed to elude my captors, and after various wanderings and strange experience made my way to the paternal home in a condition that made the Prodigal Son look like thirty cents.

That didn't abate the wandering fever in the slightest, and in order that I might not commit myself to another Walker expedition, my father consented that I should try my luck in California, and I started with his blessing and what seemed to me a liberal grub stake. I had just turned sixteen.

Instead of going to New York and taking passage from that port, I decided to travel down the Mississippi River and have a look at New Orleans, and leave on one of the various steamers that connected with the Pacific Mail at Darien.

Here an unforseen calamity very nearly upset all my plans. My money consisted of currency, issued under the auspices of the various states. A financial storm of some kind had just swept the country, and currency became legal tender only in the borders of the state of issuance. All that I could realize on my bills was barely enough to buy a steerage ticket to California. That, together with five dollars in gold coin and a revolver, comprised my earthly possessions.

At Panama we were crowded into a small steamer designed for about 400 passengers, but nearly 1,000 were crammed into it. Conditions in the steerage were appalling. Besides, the ship was under-provisioned, and we soon ran short of anything like vegetables and fruit. The purser had thriftily laid in a large private supply of oranges and bananas for sale in San Francisco. These he had divided into two caches. The hungry mob seized on one of them, located between decks, in the night, and cleaned it up to the uttermost peel. The purser knew only too well that the next night would witness the disappearance of the balance of his property. He was in despair. An inspiration seized me.

"How much will you take cash for the lot?" I asked him.

"Give me $10 for them and it's a bargain," he answered.

I fished out that lonesome $5 piece, paid it on account, and made some vague excuse about getting the other five from my bunk. I was given permission to hold a fruit auction sale on the upper deck.

Being a fruit peddler shocked my southern ideas of a gentleman's employment. Nothing but downright poverty could have driven me to it. However, I took the edge off the thing as far as possible by employing an itinerant gambler, also dead broke, to act as general salesman and orator while I took in the cash. He had a voice like a fog horn and the gall of a highwayman. He cried our wares with such success that in a few minutes the whole ship's company was engaged in mad competition to buy oranges and bananas at five for a half. It would have been just the same if I had made the price five for a dollar.

Money rolled in faster than I could count it. I could see that my chief of staff was "knocking down" on me in a shameless way, but I didn't have time to check his activities —in fact, I didn't care. In a little over an hour, the last orange and banana had vanished. I settled accounts with the purser and counted my capital. I had a little over $400 to the good, enough to make a decent start in California.

I do not tell this incident because it is noteworthy in itself. Instances were then so common of needy gentlemen who extricated themselves from the financial bog by some shift which in other days they would have thought ignoble —almost disgraceful—that this event would not be worth recalling; but in the peculiar way that destiny is worked out, it had a decisive part in directing very important matters of the future. And it has been my observation that the most impressive movements in the lives of most of us have been determined more by chance than by fixed purpose.

Among those who watched my fruit sale with interest was a gentleman named Harvey Evarts. He was a successful plumber in California and was returning from a trip to the "States," whither he had gone with a party of bankers, mine owners, and others of fortune commensurate with his own. Plumbers were not in 1857 the financial giants that they have become today. Still their stars were in the ascendant and Mr. Evarts was one of the brilliant luminaries in the sky.

This gentleman approached me after the sale. I had transferred at once from the steerage to the upper deck, as became my altered fortune, and he congratulated me in a pleasant way on my extraordinary good luck. I told him all about myself in boy fashion, and when we reached San Francisco, we had become so well acquainted that Mr. Evarts invited me to accompany him to Camptonville, then a great mining district, now off the map so far as the yellow metal goes, where he had important interests.

Placer mining was on the toboggan in 1857, when I arrived in California. All the great "bars" and gulches had been located and worked out. Very few individual strikes were made after that date. I do not know whether it was good judgment or just a case of pure "nigger luck," but at all events it happened that even in those days of declining fortune, every suggestion that Mr. Evarts gave me turned to gold. He advised me to take a chance at the head of a couple of abandoned gulches. In both cases I struck it rich enough to add $6,000 to my working capital. Again he suggested a lease of a hydraulic mine on what was known as Railroad Hill, which had been the ruin of several experienced miners. I followed his advice and, after being brought to the verge of bankruptcy, struck it rich, to my way of thinking, and cleaned up finally with $60,000 to my credit, all before my seventeenth birthday.

I visited the newly discovered Comstock Lode. Didn't like it, for deep mining seemed too slow a way of making money. Later I had a spectacular race with Jim Fair, then a hustling prospector, to locate a mining claim in Utah. But the tales of mountains gorged with wealth vanished when we got there.

Then I began to listen to a lot of mining camp talk about Mexico and its riches. California and Nevada were growing dull to my way of thinking, and I turned my thoughts to the land of Montezuma.

CHAPTER II

My Experience in Mexico

How Luck Again Brought Me Fortune

All the early gold seekers of California had some knowledge of Mexico. The great argosies of the Pacific Mail Steamship Company stopped at various points, such as Acapulco, Manzanillo, and sometimes at Mazatlán. Thus the passengers gained a sort of hurricane deck impression of the Latin nation to the southward. But it extended no further than these glimpses of the coast. A veil of profound mystery and romance shut out a view of the vast interior. Only, we knew that it was immensely rich in precious metals, but so utterly lawless and overrun with bandits that nothing short of a standing army could protect an investment.

Thus none of the adventurous pioneers attempted to explore and prospect the west coast of Mexico, which later poured its hundreds of millions into California. I may be mistaken, but I have a strong impression that I was the first of a long line of miners who went from San Francisco to Mexico and laid the foundation there for mighty fortunes.

Very much like Jason, when he pushed his classic junk from Greece, I started on my ventures in Mexico. I bought a small trading vessel, hired an excellent crew, several of whom spoke Spanish, took very little money along, but a large cargo of goods suitable to the wants of the country. In other words, I figured to make the expedition finance itself. In this I was fairly successful. After sailing up the Gulf of California and stopping at various ports, we arrived at Mazatlán, my original objective point, my cargo sold out.

There was a small American colony at Mazatlán and several groups of foreigners of other nationalities, all of them of the trading class. When I suggested a prospecting expedition into the interior, they assured me it was little better than suicide; that the country was in the absolute possession of outlaws of the most desperate type, and that a prospector's life would not be worth ten cents among them.

But I met a Mexican gentleman by the name of Don Miguel Paredis, who told me a very different story. He said that the dangers were grossly exaggerated—that there was really little to fear for anyone who understood the people. As a guaranty of good faith he offered to go with me, for at the time he happened to be broke—not an unusual incident in the life of a Mexican gentleman. Moreover, he promised to lead me to a mine of fabulous riches, in the mountains of Durango, about two hundred miles from Mazatlán. So we set out with a complete mining outfit, powder, steel, tools, general equipment and provisions for six months.

Don Miguel certainly understood his business. We really were in no more real danger than if we had been traveling through one of the New England states. We did meet some uncommonly tough-looking citizens, armed to the teeth, but Don Miguel always rode forward to meet them, handed out some specious palaver in Spanish, whereupon the whole party would disembark from their mules or horses, embrace each other on the trail, pass around some more palaver, and part with mutual esteem. The Don was a marvel as a peacemaker and I might add that for genuine good fellowship and clean dealing in all respects, he was one of the finest men of any nation I have met in a long life.

Finally, we reached his mine. This was known for years

as the Guadaloupe de los Angeles mine. He hadn't exaggerated its riches, hadn't told half the truth. The vein ran straight up the almost perpendicular face of a narrow gorge. It was merely a case of breaking down the ore as in an open cut. There were no shafts, tunnels, drifts, and winzes that take the heart out of quartz mining as a rule. And the ore was so rich that with careful sorting it was possible to make large cargoes average $500 or $600 a ton.

We never attempted to "beneficiate" or reduce the ore on the spot. Don Miguel was altogether too shrewd for that. Had bullion trains gone through the mountains from our camp, it would have taken a standing army to protect them. We simply bought mules and burros, loaded them with rock that no bandit wanted, though it was worth perhaps five hundred dollars a ton. It very seldom failed to reach the seaboard, where there were crude reduction works and plenty of purchasers of ore.

Even our inbound pack trains of costly supplies were unmolested. Don Miguel was forever practicing diplomacy. If a robber appeared at our hacienda, he was received like a friend and brother, had the best of everything, couldn't say *mas vino* too often, was handed a liberal "gratification" or tip and limitless *felicidades* on his departure. By the exercise of these arts, the management became so popular that on several occasions our pack trains were actually protected by professional bandits against marauding amateurs.

We never had a bit of trouble in our camp with the large number of people assembled there. This also was due to Don Miguel's forethought and knowledge of his people. As far as I can recollect, give the average Mexican plenty of grub, plenty of music, plenty of dancing, a little cheap finery in dress, and the rest of the world can wag on as it will, for aught he cares. He does not take kindly to abstractions, doesn't worry over his "wrongs," has no inclination to reorganize society; only wants to be let alone to enjoy the good things of life according to his own simple plan. And when you get down to brass tacks, his is not a bad philosophy, after all.

Don Miguel arranged it so that our little army of employees never had time to meditate mischief. He bought them all kinds of musical instruments, including a brass band on which they became proficient in a wonderfully short time. Every night there was a *baile* in the plaza at which the people danced till they fell from exhaustion. He offered cash prizes—mighty stiff ones—for the best dancers, male and female—the choice to be determined among themselves by a plebiscite or by select committee. Also, on Sundays, we had a bull fight. It wasn't of the sanguinary description; the bulls weren't killed, but were thriftily kept in cold storage to fight another day. It made a satisfactory sport for the people, and was also inexpensive. Added to this, we paid high wages in hard cash and kept in stock at our store an assortment of articles for personal decoration at prices that were highly profitable but not prohibitive.

Thus our enterprise became a big success from every standpoint. At a time when nearly all the mines in the Sierra Madre were closed down—practically abandoned—we were swinging along under a full head of steam, without the slightest interruption, with the general good will of all with whom we came in contact. Besides, we were making money at a rate sufficient to turn one's brain. I doubt if ever such a return was made on the trifling sum invested. There had been no development expense. The mine paid from the very day we began to operate it.

While I was the "capitalist" and owned, by our agreement, two-thirds of the property, I allowed Don Miguel an absolute free hand in all matters of policy; wherein I showed a wisdom superior to my years. And I followed his advice in one matter so important that I must mention it for the general good of mankind.

The women of the Mexican Sierra are remarkable for their physical charms. There were many real beauties resident in our camp—*simpáticas,* they used to call them—which doesn't mean "sympathetics," but "good lookers." Now, I have always believed that a good-looking woman was made to be looked at, to be admired; otherwise, wherefore was she created? Down in Mexico I could no more fail to notice a *simpática* as she passed by than I could close my eyes to the beauties of nature.

Observing which, Don Miguel gave me a piece of advice which every reader of this chapter who may happen to visit Mexico should write down for future reference.

"Leave our women alone," he counseled me. "They are romantic, soft-hearted, and will meet you halfway, but no matter how innocent your intercourse, it will rouse jealousy, ill will and serious danger. Nearly all the foreigners who get into trouble in Mexico can trace it to this source."

I realized the truth of this later when a young friend of mine called Eaton, who was a fine fellow but an ardent imitator of Lothario the Gay, was shot down in a lonely spot, jealousy being the evident motive.

In the fall of 1860, I returned to San Francisco, as I thought for a brief trip. Just to show myself, in fact. Among other things, I brought a few tons of ore that sold for $3,000 a ton, the sight of which made the town delirious. I found that my fame, or rather various romances, had preceded me. I wasn't quite twenty, couldn't vote, couldn't make a contract, yet I had over a quarter of a million in hard cash to my credit in bank, and my mine in Mexico was worth a million more. These were the actual facts, which were exaggerated and distorted beyond all resemblance to the truth. My wealth was at least quadrupled, and I was dragged through a series of bloodcurdling experiences in Mexico without a parallel in fiction.

Thus, you can see how the orange and banana sale incident set the wheels of fate revolving. If I had come to California with sufficient money, I would have made some kind of a blind stagger at luck, thrown up the sponge in disgust after a few months, and written to my father for a remittance to come home.

As it was, I quietly took rank with the great figures of the state before I had reached my majority, and became a leading actor in an unwritten page of history when the destinies of California hung by the veriest thread.

Chapter III

Story of Southern Plan to Make California Secede from the Union Is Told for First Time

Narrator Describes His Invitation into Band of 30, Which Planned to Organize Republic of Pacific

I HAD BARELY REACHED San Francisco when the election of 1860 took place, resulting in the choice of Abraham Lincoln as president of the United States. All through the South this was accepted as the signal for a civil contest. The work of organization went ahead with feverish haste, and long before the inauguration of the new President, the authority of the federal government was paralyzed in most of the slave states.

The attitude of California was a matter of supreme moment, not understood, however, at the time. Had this isolated state on the Pacific joined the Confederate States, it would have complicated the problems of war profoundly. With the city of San Francisco and its then impregnable fortifications in Confederate hands, the outward flow of gold, on which the Union cause depended in large measure, would have ceased, as a stream of water is shut off by turning a faucet. It was the easiest thing in the world to open and maintain connection through savage Arizona into Texas, one of the strongholds of the South. It does not need a military expert to figure out what a vital advantage to the Confederacy the control of the Pacific would have proved.

History relates in a few brief words how the secession movement here was extinguished by a wild outburst of patriotism. I am now going to relate for the first time the inside story of the well-planned effort to carry California out of the Union and by what a narrow margin it finally failed of accomplishment when success was absolutely secured.

I was young, hot-headed, and filled with the bitter sectional feeling that was more intense in the border states than in the states farther north or south. It would have been hard to find a more reckless secessionist than myself. I moved among my own people, got off all sorts of wild talk about spending the last dollar of my money, and my life, if need be, to resist the tyrant's yoke, and so forth, and was actually about to leave for my home in Kentucky to be ready for the impending struggle, when a quiet tip was given me that more important work was cut out where I was. My exaggerated wealth and the irresponsible stories of my Mexican exploits made me an actor in a great, silent drama, despite my years and boyish look.

One afternoon I was told to be at the house of a well-known Southern sympathizer at nine o'clock in the evening. It was well apart from other buildings, with entrances in several directions. The gentleman who owned it lived alone, with only Asiatic attendants, who understood little English and cared less for what was going on. A soft-footed domestic opened the door, took my card, and presently I was ushered into a large room where a number of gentlemen, most of them young but well established, were seated at a long table. I recognized among them leading men of San Francisco of various walks of life.

The spokesman, a great man of affairs, told me that I was trusted, that I had been selected as one to lead in an affair of great peril, an enterprise on which the future of the South might depend, and asked me if I were ready to risk life and fortune on the turn. I answered with an eagerness that satisfied my hearers and took an oath, of which I have a copy, reading as follows:

> Do you, in the presence of Almighty God, swear that what I may this night say to or show you shall be kept secret and sacred, and that you will not by hint, action or word reveal the same to any living being, so help me God?

The answer, of course, being an affirmative, I repeated after the spokesman the following objuration:

> Having been brought to this room for the purpose of having a secret confided to me and believing that to divulge such secret would imperil the lives of certain Southern men as well as injure the cause of the Southern states, I do solemnly swear in the name of the Southern states, within whose limits I was born and reared, that I will never, by word, sign or deed, hint at or divulge what I may hear tonight. Not to my dearest friend, not to the wife of my bosom will I communicate the nature of the secret. I hold myself pledged, by all I hold dear in heaven or on earth, by God and my country, by my honor as a Southern gentleman, to keep inviolate the trust reposed in me. I swear that no consideration of property or friendship shall influence my secrecy, and may I meet at the hands of those I betray, the vengeance due to a traitor, if I prove recreant to this my solemn obligation. So help me God, as I prove true.

This oath was committed to memory by every member. At subsequent meetings it was solemnly recited by all, standing and with right hand uplifted, before proceeding to further business. Several years afterwards, while it was still fresh in my recollection, I set it down in writing and preserved it to the present day. Thus I became one of a society of thirty members, pledged to carry California out of the Union.

I might say here, in parenthesis, that I have long been a reconstructed "rebel." The old flag floats over my home on every national holiday and also on Labor Day, for I take an interest in the ideas it represents. I am mighty glad now that my efforts to disrupt the Union failed and still gladder because it has been my good fortune to see the awful heritage of hate that so long divided two brave and generous people die out and disappear.

The Southern mind has a wonderful capacity for secret organization and for conducting operations on a vast scale behind a screen of impenetrable mystery. This had a fine illustration in the workings of the Ku Klux Klan in Reconstruction days, which destroyed carpetbag rule and Negro supremacy in the South and restored the government of the white race. The operations of the committee of thirty of which I became a member demonstrated the same peculiar trait.

The organization was simplicity itself. We were under the absolute orders of a member whom we called "General." He called the meetings, by word of mouth, passed by one of the members. Anything in the way of writing was burned before the meeting broke up. The General received the large contributions in private, never drew a check, settled all accounts in gold coin, and accounted to himself for the expenditure.

Each member was responsible for the organization of a fighting force of say a hundred men. This was not difficult. California at that period abounded with reckless human material—ex-veterans of the Mexican War, ex-filibusters, ex-Indian fighters, all eager to engage in any undertaking that promised adventure and profit. Each member selected a trusty agent, or captain devoted to the cause of the South, simply told him to gather a body of picked men, for whose equipment and pay he would be responsible, said nothing of the service intended, possibly left the impression that a filibuster expedition was in the wind. These various bands were scattered in out-of-way places around the bay, ostensibly engaged in some peaceful occupation, such as chopping wood, fishing, or the like, but in reality waiting for the word to act. Each member of the committee kept his own counsel. Only the General knew the location of the various detachments.

Our plans were to paralyze all organized resistance by a simultaneous attack. The Federal army was little more than a shadow. About two hundred soldiers were at Fort Point, less than a hundred at Alcatraz, and a handful at Mare Island and at the arsenal at Benicia, where 30,000 stand of arms were stored. We proposed to carry these strongholds by a night attack and also seize the arsenals of the militia at San Francisco. With this abounding military equipment, we proposed to organize an army of Southern sympathizers, sufficient in number to beat down any unarmed resistance.

All of which may seem chimerical at this late day, but then, take my word, it was an opportunity absolutely within our grasp. At least 30 per cent of the population of California was from the South. The large foreign element was either neutral or had Southern leanings. We had already, under practical discipline, a body of the finest fighting men in the world, far more than enough to take the initial step with a certainty of success.

And those who might have offered an effective resistance were lulled in fancied security or indifferent. It is easy to talk now, half a century after the event, but in 1860 the ties that bound the Pacific to the government at Washington were nowhere very strong. The relations meant an enormous loss to California. For all the immense tribute paid, the meager returns consisted of a few public buildings and public works. Besides, thousands were tired of being ruled from a distance of thousands of miles. The "Republic of the Pacific," that we intended to organize as a preliminary, would have been well received by many who later were most clamorous in the support of the federal government.

Everything was in readiness by the middle of January, 1861. It only remained to strike the blow.

Chapter IV

Southern General, Albert Sidney Johnston, Played Important Part in Preventing Organized Revolt for Secession

Discovery of Comstock Lode with Its Vast Hoard of Gold Another Factor in Keeping This State in the Union

General Albert Sidney Johnston was in command of the military department of the Pacific. He had graduated from West Point in 1826 and saw seven years of active service on the frontier, especially in the famous Black Hawk War. He resigned from the service on account of his wife's failing health, and settled in Texas. On the uprising against Mexican rule, he had enlisted as a private soldier in the army of his new country, but through the compelling force of genius soon became commander-in-chief of the Republic's forces. At the time of the annexation of Texas, he was its secretary of war. When the war with Mexico broke out, he offered his services to the United States, fought in many of the severe engagements, rose to the rank of general, and was sent to Utah to suppress what was known as the "Mormon Rebellion," which he accomplished with firmness and tact. In January, 1861, he was placed in command of the Department of the Pacific.

Johnston was born in Kentucky, but he always in later years spoke of and considered Texas his state. Thus he had a double bond of sympathy for the South. This was the man who had the fate of California absolutely in his hands. No one doubted the drift of his inclinations. No one who knew the man and his exacting sense of honor doubted his absolute loyalty to any trust.

In all of our deliberations, General Johnston only figured as a factor to be taken by surprise and subdued with force. We wished him well, hoped he might not suffer in the brief struggle, but nobody dreamed for an instant that his integrity as a commander-in-chief of the army could be tampered with.

One of the most brilliant members of the early San Francisco bar was Edmond Randolph. He was a man of rare talents and great personal charm. Born in Virginia, a member of the famous Randolph family, he was naturally an outspoken advocate of the South. He was one of our committee, and on terms of social and professional intimacy with every one of Southern leanings. He was on the closest terms with General Johnston, and there is hardly a doubt that, purely on his own motion, he approached the General with some kind of a questionable proposition. What happened at that interview no man knows, but Johnston's answer made Randolph stark crazy. He indulged in all kinds of loose, unbridled talk, told several of our committee that he had seen Johnston, that the cause was lost, and otherwise, in many ways, exhibited an incredible indiscretion that might easily have been fatal to our cause. No amount of warning was able to silence his unbalanced tongue.

This situation was discussed at several meetings and finally it was decided that a committee of three should visit General Johnston in a social way, not to commit further folly by any intimation or suggestion, but to gather, if possible, some serviceable hints for future use. I had become prominent in council through my zeal and discretion, and to my great joy I was named as one of the three.

I will never forget that meeting. We were ushered into the presence of General Albert Sidney Johnston. He was a blond giant of a man with a mass of heavy yellow hair, untouched by age, although he was nearing sixty. He had the nobility of bearing that marks a great leader of men, and it seemed to my youthful imagination that I was looking at some superman of ancient history, like Hannibal or Caesar, come to life again.

He bade us courteously to be seated. "Before we go further," he said, in a matter-of-fact, off-hand way, "There is something I want to mention. I have heard foolish talk about an attempt to seize the strongholds of the government under my charge. Knowing this, I have prepared for emergencies, and will defend the property of the United States with every resource at my command, and with the last drop of blood in my body. Tell that to all our Southern friends."

Whether it was a direct hint to us, I know not. We sat there like a lot of petrified stoten-bottles. Then, in an easy way, he launched into a general conversation, in which we joined as best we might. After an hour, we departed. We had learned a lot, but not what we wished to know.

Of course the foreknowledge and inflexible stand of General Johnston was a body blow and facer combined. There was another very disturbing factor—the Comstock lode.

While we were deliberating, the marvelous mineral treasure house began to open up new stores of wealth. Speculation was enormous. The opportunity for making money seemed without limit. Many of the committee were deeply interested.

Now it had been determined absolutely from the outset that our ambitions were to be bounded by the easily defended Sierra. We knew enough about strategy to under-

ALBERT SIDNEY JOHNSTON

Commanding the Military District of the Pacific in 1861

stand that it would be simple madness to cross the mountains. That meant, of course, the abandonment of Nevada.

This had been accepted with resignation when the great mines were considered played out. But when it became apparent that the surface had been barely scratched and that secession might mean the casting aside of wealth beyond the dreams of avarice, then patriotism and self-interest surely had a lively tussle. If Nevada could have been carried out of the Union along with California, I am almost certain that the story of those times would have been widely different. We certainly had the organized forces to carry out our plans.

That's the only way I can size up what followed. The meetings began to lack snap and enthusiasm. Just when we should have been active and resolute, something always hung fire.

The last night we met, the face of our General was careworn. After the usual oath, he addressed the committee. It was plain, he said, that the members were no longer of one mind. The time had now come for definite action, one way or another. He proposed to take a secret ballot that would be conclusive.

The word "yes" was written on thirty slips of paper; likewise the word "no." The slips were jumbled up together and were placed alongside of a hat in a recess of the large room. Each member stepped forward and dropped a slip in the hat. "Yes" meant action; "no," disbandment. When all had voted, the General took the hat, opened the ballots and tallied them; then threw everything in the fire. "I have to announce," he said, "that the majority have voted 'no.' I therefore direct that all our forces be dispersed and declare this committee adjourned without day."

Not another word was spoken. One by one the members departed. All I can say is that they kept their secret well.

Two days later, all the various bands had been paid off and dispersed. The "great conspiracy," if you wish to call it so, had vanished into the vast, silent limbo of the past.

Only the General knew the extent of the disbursements. My own impression is that they far exceeded a million dollars. I contributed $100,000 myself, which, of course, was an incident of the financial recklessness of youth.

Many of the committee rose to great social and public prominence. The "General" died not so long ago, full of years and honors.

Besides myself, there is one survivor, whose name would surprise the nation.

(Since the above was first printed, this survivor has died.)

CHAPTER V

Randolph Betrayed Conspiracy for Revolt in California, and Wrote Letter to Lincoln that Caused Johnston's Removal

I COULD NOT CLOSE this phase of the story without further reference to Edmond Randolph, for I sincerely want to set him right. I said he went mad. Everything later proved it. He not only committed the gravest indiscretions, but in addition, he, a Southern man, with a couple of centuries of Southern traditions behind him, actually wrote a letter to President Lincoln warning him of a vast conspiracy to carry California out of the Union and questioning the trustworthiness of General Johnston. Nothing but downright lunacy could have inspired the act. This was sent to President Lincoln by pony express and reached him just about the day of his inauguration. The story has been often printed before or I would not revive it now. Its accuracy has indeed been questioned by Randolph's friends. I am inclined to believe it true.

As a consequence General E. V. Sumner was sent on a tug from New York with sealed orders and placed on board a Pacific Mail steamer in midocean. On the steamer the orders were opened. They directed him to proceed to San Francisco and relieve General Johnston of the command of the Department of the Pacific. History relates further that General Sumner was taken from the steamer by a government vessel outside the Golden Gate, hurried to Alcatraz, where General Johnston had headquarters, and, in a sensational manner, relieved him of his command.

The latter part is purest fiction. General Johnston never had headquarters on Alcatraz. He lived with his family on Rincon Hill, near the residence of Louis Garnett. Sumner arrived in San Francisco on the steamer, publicly, like anyone else. General Johnston, informed of his arrival, at once arranged for a conference, and the two met in perfect amity at the old army headquarters, located on Bush street, if I recollect aright. The transfer of authority took place the next day. There are abundant living witnesses to these facts. General Johnston's resignation was in President Lincoln's hands long before Sumner reached California and the same was accepted a few weeks later.

One of General Sumner's first acts was to order arms from the arsenal and organize patriotic citizens for an expected crisis. But they were simply fighting windmills. The real crisis had disappeared of itself two months before, through General Johnston's firmness—and the Comstock lode.

As a further proof of Randolph's madness, he straightway developed into an outspoken, rabid secessionist, made speeches of the most inflammatory nature, and it was highly significant that he escaped imprisonment in Alcatraz. He died within the year, a physical and mental wreck. In my humble judgment he deserves sincere pity, not blame.

That someone of important station wrote a mysterious letter to President Lincoln which caused the retirement of General Johnston is beyond dispute.

One of the versions of the story has never been published, to my knowledge. In 1880, when Mr. Justice Field was candidate for President, he flooded the South with literature concerning his friendship for that section, as evidenced by various decisions of the United States Supreme Court in the dark days of Reconstruction. In the North, principally among the Grand Army, a pamphlet was circulated to the effect that he had saved California to the Union by a timely letter to President Lincoln, which resulted in General Sumner's hasty mission. Whether it was authorized by Judge Field, I do not know. But it fell into the hands of the Southern leaders and doomed his candidacy in the section where he counted on support. Not at all because he had saved the Union, but because of the implied aspersion on the memory of one who will ever be dear to the South—a gentleman of unimpeachable honor, a great soldier who died a soldier's death, fighting for the Lost Cause.

After he resigned, General Johnston earnestly advised many Southerners, some of them still alive, to do nothing that would bring war to California. "If you want to fight, go South," was his constant counsel to all. Many followed his advice. Hundreds, perhaps thousands, of them were cut off by Indians in Arizona, where the savages had full swing, all the frontier army posts having been abandoned. General Johnston stayed in California till his state—Texas—seceded. Then with a few followers he traversed the savage wilderness and after many adventures reached the South.

There is a rather pathetic sidelight to the story that illustrates the simple devotion of the old-time slaves to their white masters. General Johnston had freed all his slaves before he came to California. One of them, called "Rand," brief for "Randolph"—he had no other name—followed him as a body-servant to the Pacific. When Johnston left for the South, he ordered "Rand" to stay behind. He was a famous cook and could have commanded big wages in a high-class restaurant. But the faithful body-servant would not be denied. He fought his way with his former master through the Apaches of Arizona and was with him at Shiloh when he died. He hung over the dead body of the fallen leader in a wild passion of primitive grief.

Later some hundred colored body-servants of General

Johnston's appeared at various parts of the South. The real "Rand" settled in Louisville, where he was an object of solicitous regard on the part of the Johnston family and others of the old regime.

"Rand" proved himself no less great in peace than war, for he married a widow with seven children, an act that needed moral courage of the highest sort. His career was somewhat checkered, but he was always well looked after, and "looking after" "Rand" was often quite a job.

He became something of a character in the border city; resolutely declined to be "reconstructed" and remained an unrepentant rebel to the last. He was very bitter in his talk about the "poor white trash" of the North. When he uncorked the vials of his wrath, he called his adversary an "abolitionist" as the last word of scorn.

In his final illness tender Southern hands smoothed his way into the hereafter. Mrs. H. P. Hepburn of Louisville, once of San Francisco, was present when the curtain rang down on "Rand." He raised his feeble head and said: "I'se 'gwan to meet ole Marse Johnston," then sunk back on the rough pillow, closed his eyes and died.

CHAPTER VI

Perilous Trip Across Mexico and Voyage on Blockade Runner Enter into Narrator's Experiences on Visit to Jefferson Davis

Southerners in California Form Plan to Intercept Gold Shipments on Pacific Mail Liners from San Francisco to Capital

I WAS BROKEN-HEARTED at the turn of affairs in California. Needless to say, I was one of those who voted "yes" on the memorable night when the committee disbanded. The actions of General Sumner, which were needlessly severe and autocratic, tended to make the tension more severe. Just for some idle expression of sympathy for the South, all sorts of really inoffensive people were clapped into Alcatraz and subjected to indignity and loss. President Lincoln later on realized that Sumner was only making matters worse and sent General Rice to relieve him, who at once adopted a policy more pacific and wise.

But this is no part of the story. The idea of interrupting the gold shipments by the Pacific Mail, very essential to the government at Washington, again took form. This was to be effected by seizure on the high seas. A number of prominent men were interested and I was requested to become one. I had no stomach for downright piracy, though ready for any risk. I stipulated that I must first receive a regular commission from the Confederate Navy. This being agreed to, the sum of $250,000 was subscribed, of which $50,000 was mine.

In company with H. T. Templeton, a well-known Californian, later a familiar of the Crocker family, we traveled by steamer to Acapulco. Mexico was then in an uproar over the threatened French invasion. The American consul, a son of John A. Suter [Sutter?] advised us that it was little short of madness to cross the country to Mexico City, which we gave as our destination. But Templeton was brave as a lion and I was young, reckless, and confident in my luck. Heavily armed, with a single guide, who, by the way, fled in terror at the first sight of danger, we set out on a venturesome journey.

That trip would make some story by itself. We had several pitched battles with small bands of *"ladrones"* or robbers. Once both our horses were shot from under us. My previously acquired knowledge of Spanish stood us in good stead in securing fresh equipment, knowledge of the way, and sometimes hospitality and shelter. Finally, after great hardships and danger, we reached Mexico City, and thence proceeded without incident to Vera Cruz, which was a sort of rendezvous for blockade runners. Here Templeton and I parted company with mutual regrets. He took a ship for New York and returned to California. I boarded a blockade runner and during a rainy night we slipped past the Federal warships into Charleston.

I had no difficulty in reaching Richmond, Virginia, the Confederate capital. It was a vast, hustling, military camp. Troops were marching and countermarching, officers on horseback dashing to and fro on mysterious missions and everywhere the atmosphere of war.

It was a couple of days before I saw President Jefferson Davis. I laid my plans before him fully, to his great interest, and later we had several interviews. He did not come to a swift conclusion. To my way of thinking at the time, he was over-deliberate in making up his mind. That was a youthful illusion. I think of him now as a very great man, lacking only one thing—luck.

He fully realized the importance of shutting off the great

gold shipments to the East from California. President Davis said it would be more important than many victories in the field. At the same time, he saw grave difficulties in the way. He did not believe that a vessel could be outfitted for the purpose in any of the Pacific ports without arousing suspicion, disclosure, and capture. He warned me that my associates and myself were taking an awful risk, almost sure to result in ultimate disaster. Moreover, he was uncertain whether under any circumstances the enterprise could be justified under international law and whether the proceeding would not fall under the head of piracy, against which he resolutely set his face.

All these questions were submitted to one of his cabinet officers, Judah P. Benjamin. Mr. Benjamin was of Jewish ancestry and one of the ablest men who guided the way of the Confederacy. After the general breakup, he escaped to England, became a leader of the bar of London, counsel to the Queen, and won the highest honors of his profession before he died. This distinguished gentleman examined with great care the questions involved, particularly on the piracy point, and he gave an opinion that it would be entirely within the scope of international law to equip and sail a vessel out of any port of the United States provided no overt act against commerce were committed before a foreign port was reached, letters of marque exhibited there, and the open purpose of those in command declared. So for what followed I had at least the advice of eminent counsel, and I still believe that the advice was absolutely sound.

In due course of time I received a commission as a captain in the Confederate Navy. I had never been on a man-of-war in my life, but that made no difference. A fresh-water naval hero may be as good as the salt-water kind. Also, I received letters of marque in blank, the names to be filled in when the vessel reached a foreign port. Besides that I was intrusted with quite a bundle of mail, addressed to leading Southerners in California and doubtless of a highly compromising character.

This literary consignment nearly got not only myself but many other people into a peck of trouble, which I might as well tell of now, although it is somewhat ahead of my story. Returning to California, liking not the route through Mexico, I had the blockade runner land me at Aspinwall, where I joined the passengers of a Pacific Mail liner and embarked at Panama for the run north. As we were approaching San Francisco, I became uneasy about my documents, fearing that enough of my movements might be known to cause a close personal search.

On board the steamer was a lady long famous in California, Mrs. Charles S. Fairfax. Her husband was the lineal Lord Fairfax of the British peerage. She was the niece of John C. Calhoun, a woman of great beauty, wit, and resourcefulness and an intense Southern sympathizer. We became rather confidential on the way up and I told her about the package and my fears.

"Why, what stupid fools men are, anyhow," she laughed, "give that package to me and set your mind at rest." The suggestion looked good, for, of course, I could assume responsibility if the documents were found. That night Mrs. Fairfax left her door just a bit ajar, and as I passed it, something was slipped to her. No one saw the transfer.

When we reached San Francisco what I feared came true. Not alone my luggage, but my person were subjected to a search that hardly overlooked my soul. While I was in the hands of the minions of the law, who seemed sadly disappointed over their fruitless quest, Mrs. Fairfax swept by in her stately way; all the same I seemed to catch a twinkle of humor in her eye.

Two days later, the lady handed me the package. The

seals were broken, but the contents intact. "You gave me a lot of bother," said the lady, "I had to sit up all night sewing these wretched papers in my dress. What was worse still, I never dared to change it. Just imagine what the other women thought of me."

I passed the letters around to various leading lawyers, bankers, financiers, and so on. Without mentioning any names, I told them how near they came to falling into Federal hands. Many a cheek paled and jaw dropped as they heard the story.

We have been told much of what women did for the North, very little of what the women did for the South. That is a noble and inspiring story that remains to be told.

But to return to Richmond. The Confederate cause seemed at its zenith. Everywhere was abounding confidence in the final result. And now came a whisper that a great battle would soon be fought that ought to be decisive. I was eager to see something of the war game and with letters from the Secretary of War, hurried westward, arriving at Corinth, Miss., on April 4, 1862. Here a small Confederate army was assembled under the same Albert Sidney Johnston, not exceeding 5,000 men. Nine miles away, General Grant was encamped at Shiloh with 35,000 men, confidently awaiting the arrival of General Buell with 30,000 more, to begin the invasion of the South.

At the risk of criticism by experts I am going to tell briefly what a great, old-fashioned battle seemed like to a raw looker-on.

CHAPTER VII

The Great Battle of Shiloh and the South's Irreparable Loss in the Death of General Johnston

WAR, FIFTY YEARS AGO, was bad enough, but it wasn't the plain, cold-blooded deviltry that it is today. When men met face to face and leaders led, in fact as well as theory, I can understand the inspiration, the enthusiasm, the wild love of glory, that invited the best blood to a military life. But now, when victories are to be won by pressing buttons, switching on or off electric currents, or dropping bombs from the sky on the heads of helpless women and children, while it may attract those of a mechanical turn of mind, it has ceased to be a business that should interest a gentleman.

My recollection is of the old fighting days. I said that when I arrived in Corinth on April 4, 1862, not more than five thousand men were assembled there. But all that night and the next day troop trains were unloading enormous reinforcements and some were arriving by forced marches on foot. By the night of April 5, between twenty-five and thirty thousand soldiers were in camp, the flower of the fighting army of the South. General Albert Sidney Johnston, with his heroic figure and magnetic presence, roused the men to a height of martial exultation very hard to describe. Everyone knew that a great battle was impending. Most of them guessed that the morrow would be the day. But they hardly seemed able to wait. They were like war dogs tugging at the leash, confident in themselves, confident in their cause. One would have thought they were bound for a holiday excursion instead of a death grapple from which many would never emerge.

Very much to my disappointment, I was assigned to the staff of General Beauregard, second in command. I had hoped to be with General Johnston, where the fighting would be the fiercest. Nevertheless, I had enough.

The troops retired at an early hour on the night of April 5. But in the darkness flitted shadows of alert men, making busy preparations for a great event. At two o'clock in the morning, the troops were roused from their sleep, had hasty refreshment in the darkness, and then fell in, company after company, like so much clock work, and the march to Pittsburg Landing, or Shiloh, nine miles away, began. The infantry was well in front, separated by perhaps half a mile from the artillery and more noisy equipment.

The nature of the country was admirable for a secret movement. It was well wooded, with abundant cover to screen our presence, and it seemed almost uncanny how the thousands of men marched forward with scarce noise enough to stir the early morning air. Not a word was spoken.

It was just daylight when we drove in the Federal pickets. Before us lay the army of General Grant. It seems to me that it was not more than two hundred yards away. Breakfast was being cooked, the officers and men totally off their guard. Nothing in the nature of surprise could be imagined more terrible and complete. Quick commands were given, there was a rattle of musketry, the "rebel" yell rang out—a sound that might well start the resurrection of the dead—and the next instant I saw what appeared a long line of racing apparitions in gray, with fixed bayonets, clear the intervening space and fall like a cloudburst on the men in blue.

Nothing saved the army of General Grant from utter destruction but the presence of several gunboats in the

Tennessee River. These were splendidly handled, and the fire was deadly and precise. It gave the Union forces an opportunity to recover somewhat and put up a gallant fight. Field artillery was concentrated on the gunboats. Sharpshooters climbed into near-by trees and picked off the gunners at their posts. The fire became less frequent, less precise.

Anyone could see the line of General Johnston's strategy. Grant's army was encamped on rising ground beyond the Tennessee. Behind it the ground fell off rather abruptly to a narrow plain along the river bank, beyond which there was no retreat. The object of the attack was to force the Federal line to the river bank and then drive in the wings until the Union army became a huddled mass on the low ground where it could not fight effectively, and be at the mercy of the artillery fire. Then it must either surrender or be wiped out. The first step was accomplished by the initial bayonet charge. The second required more time.

The battle raged into the afternoon. The field was covered with dead and dying, but the strategy of General Johnston was rapidly bearing fruit. The gunboats were almost silenced, the Federal columns showed apparent signs of disintegration. Another hour would have seen a total rout. General Johnston had been everywhere, the directing genius, exposing himself to needless dangers. Just in the moment of triumph, he fell headlong from his horse.

It seemed as if the news of this irreparable loss spread through the army like wildfire and caused not a demoralization, but a general pause. Beauregard took command, evidently under a great mental strain. To the surprise of many, he gave orders to retire. I heard him say: "Tomorrow we will be across the Tennessee River, or in hell."

He had another guess. Early the next morning General Buell crossed the Tennessee with thirty-five thousand fresh troops, and all day we were fighting our way back to the strong position at Corinth. The great opportunity was lost.

Thus I saw the bloodiest battle of the war, and I think the most decisive—far more so than Gettysburg. Had Johnston overwhelmed Grant at Shiloh, met Buell with an army flush with victory, with no gunboats to contend with, there might have been another tale to tell. With Tennessee liberated, Kentucky and Missouri might have joined the Confederate cause and influenced the final outcome profoundly. When I look back at the long series of mishaps and unforeseen misfortunes that seemed to haunt the Lost Cause, I cannot but conclude that God's will was there. After many years of bitter recollections, we are all of one mind—that the outcome was best for the country, and best of all for the South.

I saw General Johnston's body on the field, where he fell. The wound that caused his death was of a trifling nature. A rifle ball had cut an artery in his leg. A surgeon with a tourniquet could have stopped the hemorrhage. But he never sought assistance. He stood by his post like a true soldier, and slowly bled to death.

History has classed Johnston as a great military genius. Years after, the government of the United States erected a shaft with a suitable inscription on the spot where he fell at Shiloh. His tomb, with a noble equestrian statue, is in New Orleans. Most of his direct descendants live in California, the state that he saved from the desolation of war.

Concerning the battle of Shiloh, I have better testimony than my own. A score of years later, I met General Grant in New York. Out of an acquaintance, an intimate friendship developed. During his first financial embarrassment, of which the world never knew, I piloted him to a safe haven. Grant's genius was entirely one-sided. In matters of business, he was the veriest child. He had tied himself up in Wall Street ventures and was facing ruin when he sought my advice. I took his account to my brokers, Henry Clews & Company, where I had a balance of nearly two millions to my credit, and, by careful nursing, brought him out, not only even, but ahead. The General and I often spoke of Shiloh, and he admitted, with a soldier's frankness, that only Johnston's death saved his command. He also added that he learned a lesson in war that fateful day, the most important in his long experience.

In this era of good-will and reconciliation, when the old boys in blue and gray are meeting in comradeship on the scenes of their former struggle, why cannot someone write a trustworthy and impartial history of the great drama—the greatest of our national life—which our boys and girls may read and learn the truth? The textbooks of our schools are still deformed by a spirit of intolerance and prejudice, most unfortunate and misleading in an age that has happily outlived the bitterness that divided us in the past.

CHAPTER VIII

Nephew of Celebrated English Leader Takes Hand in Conspiracy, and Also Figures in Amusing Near-Duel

I DID NOT RETURN to California after my visit to the seat of war until late in the month of July, 1862. Everything seemed in regular shape for outfitting a privateer. But again the Comstock Lode interfered. Speculation was fast and furious. Of those who subscribed to the fund of $250,000 to carry on the enterprise only two remained steadfast, Mr. Ridgley Greathouse and myself. Greathouse was connected with some of the well-known families of the South and of California. He was a man of unusual courage and determination. We laid our heads together and decided to go ahead alone.

At this point we gained an unexpected ally. As he cuts quite a figure in this story, especially in the great diamond hoax, I might as well explain the strange way in which we met.

Mr. Alfred Rubery was a young English gentleman of fortune and culture, with the roving disposition and love of venture that was part of the make of high-strung Englishmen of his day. Traveling in the South just before the war, he had acquired an admiration for its aristocracy. Thus happened something that seemed paradoxical. Rubery was the favorite nephew of John Bright, the great English statesman and publicist. It was due to his influence and leadership among the laboring masses that England declined to interfere in favor of the Confederate States when its industries were ruined and the industrial classes starving, because the cotton staples from the South, on which they depended, were suddenly cut off. Thus, while John Bright, across the Atlantic, was resolutely upholding the North, his dear nephew in San Francisco was openly expressing sympathy for the South.

Sectional feeling at that period was so intense that the slightest word brought on a quarrel. One evening Rubery met a young officer, Lieutenant Tompkins, stationed at Fort Point, scion of a prominent family of New York. Somehow the subject of the war was broached. High words followed, and Tompkins made a remark that touched Rubery's honor. The latter simply said, "You will hear from me, sir," and left the room.

The code duello was still in full force. Though a cause of instant dismissal from the army, no officer would hesitate for a moment to refuse satisfaction to a gentleman who considered himself aggrieved. Rubery sought a friend of mine and asked him to bear his challenge. He was on the point of leaving for Oregon to attend to some of my business. For that reason he turned the young Englishman over to me.

Now, when a man chose a second, he placed his life entirely in his hands. It became at once my duty to examine certain details. The challenged party had the right to name the weapons, and I knew Tompkins to be an expert swordsman. I asked my man about his saber experience. He admitted that he had some knowledge of carving ham, but as to carving anything else he was as ignorant as a child. I tried him at pistol practice and found that, with extra good luck, at ten paces he could hit a barn.

To go into a duel under such conditions was downright madness. I told Rubery that I could not suffer him to be a chopping-block for a Yankee or to be coolly potted while he was shooting at the sun. I advised him that he must take time to practice with swords and pistols. But the Englishman would not be denied. I never saw a man so determined. He said he would rather die a thousand times than survive an unresented public insult. Having no alternative, I carried Rubery's challenge to Tompkins at Fort Point.

Lieutenant Tompkins referred me to his friend, Quartermaster Judson, whom I met without delay. I found he had little stomach for the duel, not because he or his principal were afraid, but because they dreaded dismissal from the service. He admitted that his principal was in the wrong and asked if there were any reasonable terms to adjust the difference. I told him I was instructed by my principal to accept nothing but a written retraction of the offensive language. "That is out of the question," said Judson. "We are wasting time. Let us proceed to details."

"Proceeding to details" was quite a formal function in the code. Arrangements for the slaughter of a couple of human beings were always discussed over a bottle of wine, in a spirit of friendly benevolence. Judson produced the refreshments, filled my glass, handed it to me standing, left his own unfilled and sat down.

Now, in southwestern Kentucky, where I was raised, gentlemen always drank together. To offer wine or corn juice to an equal and not partake yourself was an almost unpardonable affront. You might do that without offense to an humble dependent, but not to one of the same social rank.

I had determined that the duel should not take place and was watching for any chance to spar for time. This seemed to offer an "opening." Of course, Judson had not the most remote idea of being discourteous. But I assumed to think otherwise. I looked as indignant as possible, dashed the glass on the floor, slapped my hat on my head and left the apartment before the astonished quartermaster had time to catch his breath. A few hours later my second,

ALFRED RUBERY

Nephew of John Bright, the great British publicist

[*Reproduced from an old photograph*]

Captain Fluson, a famous duelist, waited on Judson with my challenge.

I hope no one will imagine I am bragging. I took not the slightest chance in sending the challenge and knew it very well. No man was compelled to accept a challenge without a full knowledge of the nature of his offense. If a person wanted to fight you just for his own amusement or because he disapproved of the cut of your coat, no one was expected to humor him, and a man of honor could properly refuse to consider a challenge based on trivial grounds or even kick the bearer out of doors. As soon as my second presented himself to Judson, just as I expected, he asked to be informed in what way he had given offense to Mr. Harpending. My second explained the deadly nature of the one-sided invitation to drink, according to the usages of southwestern Kentucky, whereat the quartermaster laughed and said he was ignorant of any such custom; that he had never had the remotest intention of being discourteous and asked that this explanation be given me before going further.

Of course, I had to appear immensely gratified. I wrote Judson, expressing my entire satisfaction, apologized for my own hasty conclusion, and asked him to dinner. We had a jolly sort of time and over black coffee we discussed the proposed Rubery-Tompkins duel. Both agreed it was a shame to see two fine young fellows fill each other with lead and decided to co-operate to prevent it. We managed to bring the principals together, and after a lot of diplomacy on all sides, Tompkins agreed to a written retraction of the insulting language, Rubery promising that it should never be exhibited unless he were charged with cowardice as a result of the billiard-hall incident.

Everything terminated in a dinner party and the incident was closed.

Rubery and I, thus strangely brought together, became inseparable. We were nearly of an age, both crazy for adventure, both devoted to the South. It was not, therefore, strange that I confided to him all my plans of outfitting a privateer. When he learned the details he became almost idiotic with delight. "Now, we're getting somewhere," he cried. "Let me be your associate and count me in to the limit."

That is how the nephew of John Bright became associated with Greathouse and myself in an effort to destroy the commerce of the Pacific Coast and how he came to loom largely in what was known to history as the "Chapman piracy case."

CHAPTER IX

Plan to Capture Gold Ships Develops, But Trouble Follows Engagement of Villainous-looking Pilot

The three of us—Greathouse, Rubery, and myself—now worked in unison. My first intention was to outfit in British Columbia, but an agent stationed at Vancouver was unable to find anything fit for our purpose. We negotiated for the purchase of the steamer *Otter,* owned in Oregon, but on a trial trip she failed to develop a speed much greater than that of a rowboat—not enough either to fight or run away.

While we were fretting over the delay, a small deep-water vessel came into port, after a record-breaking voyage from New York. The ship was called plain *Chapman.* Historians have seen fit to name it the *J. M. Chapman,* for what reason I am not aware. Probably it was a case of what literary folk are pleased to call "poetic license." At any rate, we considered it a serviceable craft, in default of a steam vessel. We purchased the *Chapman* from her owners at a reasonable price, as it was winter and an outbound cargo was not obtainable at that season of the year.

Our plans might as well be explained fully here. We proposed to sail the Chapman to some islands off the coast of Mexico, transform her into a fighting craft, proceed to Manzanillo, exhibit our letters of marque and my captain's commission in the Confederate Navy and then lie in wait for the first Pacific Mail liner that entered the harbor, capture her—peacefully if possible, forcibly if we must. All of this was in line with instructions. Then we proposed to equip the captured liner as a privateer and figured to intercept two more eastbound Pacific Mail steamers before the world knew what was happening, in those days of slow-traveling news. After that we proposed to let events very much take their own course. It was a wild, desperate undertaking at the best, but we were all of an age that takes little stock of risks.

Having our ship, other details followed rapidly enough. We purchased two cannons throwing a twelve-pound shot. This was arranged by a Mexican friend of mine, acting through a well-known business firm, which was entirely ignorant of the nature of the transaction. In the same way, we bought shells and solid shot and a large quantity of ammunition. In those days of adventure it was no uncommon matter for corporations or even private persons to purchase armament on a considerable scale, without comment. Often remote investments had to be protected not only with armed men, but also with a show of artillery. Our Mexican friend merely had to say that he needed the military supplies to guard a mining property in his own country. As a matter of fact, he never knew what the war material was intended for—just took it for granted that he was doing something in the line of accommodation.

Also we bought a large assortment of small arms, rifles, revolvers, and cutlasses. Everything was heavily boxed and marked "machinery." We laid in also, to avoid suspicion, a small line of general goods of a kind salable in a Mexican port, and an extra supply of provisions.

We engaged an ordinary crew of able seamen and without much difficulty selected twenty picked men—all from the South, of proved and desperate courage. These were to constitute our working force. They were not known to each other, did not even know the nature of the service—further than that it meant fighting and plenty of it—somewhere in Mexico.

All our plans were perfected. It only remained to secure

a navigator who could be implicitly trusted. Men of the South did not have much practical experience in seamanship. Several of our confidential friends scoured the town for a suitable person for this all-important post.

Finally a man was brought to me by the name of Wm. Law, guaranteed to be a competent navigator familiar with the Mexican coast and a Southern sympathizer. He was the possessor of a sinister, villainous mug, looked capable of any crime, and all in all was the most repulsive reptile in appearance that I ever set eyes on. From the moment I saw him, I was filled with distrust. After a short general conversation I dismissed him and told his vouchers that I could put no faith in such an ill-omened-looking character. But time was pressing. No one else showed up, and after further guaranties, Greathouse, Rubery, and myself saw Law again and frankly gave him a general outline of our plans. He accepted the responsibility with a well-feigned eagerness; his tough-looking face seemed lighted with a sort of demoniac exultation. There was still another who shared our confidence to some extent, Libby, the sailing master of the *Chapman*.

Everything was now ready to launch the enterprise. Our clearance papers were secured from the customhouse with a readiness that might have suggested a suspicion to more alert minds. The *Chapman* was duly certified to sail for Manzanillo with a cargo of machinery and mixed merchandise.

It was on the night of March 14. Greathouse and Law were to be on board at ten o'clock. Rubery and I stationed ourselves in a dark alley behind the old American Exchange Hotel. One by one, our fighting men assembled silently, by prearrangement. The night was dark, the sky overcast. We divided into three squads to avoid attention, slipped through the dimly lighted streets past roaring saloons and sailor boarding houses, and reached an unfrequented part of the water front unnoticed, where the privateer was moored.

Everything thus far had gone so smoothly that Rubery and I were exultant. The wind, too, was propitious. We figured to sail without delay, pass Fort Point in the dark, and be beyond the horizon before the morning broke. We scrambled aboard the *Chapman*. Greathouse was pacing the deck in agitation. Law was not there.

I experienced a shock such as a man receives when a bucket of ice water is emptied on him in his sleep. The suggestion of treachery could not be avoided. We cast loose from the wharf and anchored in the stream. But we were helpless. We could not sail without our navigator. We had nothing to do but wait.

We scanned the bay for an approaching boat, but the dark waters answered not. At two o'clock we turned in for a much needed rest. We left a trusty man as a lookout, with orders to waken us at five o'clock if nothing happened before. We still had a lingering hope that Law might appear in season to carry out our plans. And soon, as the hours glided by, the *Chapman* rocked us to sleep.

CHAPTER X

We Wake to Find Warship Near and Boat Filled with Police Approaching

Somebody else slumbered on board the *Chapman* that night besides the men below. Morpheus evidently got a strangle hold on our vigilant sentinel, from what followed. I was wakened by a shake and a startled cry from the lookout. I sprang hastily to the deck.

It was broad daylight. A couple of hundred yards away I looked into the trained guns of the U. S. warship *Cyane*. Several boatloads of officers and marines were just starting from her in our direction. A hasty look also revealed a tugboat making for us from the waterfront, filled with San Francisco cops, headed by I. W. Lees.

Of course, even had we been prepared, resistance would have meant suicide, for the gunners of the *Cyane* stood waiting orders to blow us out of the water. I rushed down to the cabin, jerked Rubery and Greathouse from their bunks, and after a brief word of explanation, we proceeded to destroy as many incriminating papers as possible. We made a hasty bonfire on the cabin floor, burned a number of documents that might not have looked well if read in open court, tore into little bits and scattered the fragments of other documents that resisted a quick fire, and made a cleanup in general. Smoke was streaming up the gangway when naval officers and policemen swarmed on board. Someone yelled, "They've fired the powder magazine." This made a diversion and gained a little more time. Nevertheless, out of the destruction, Captain Lees gathered together the scraps and by piecing them together and guessing at the missing parts, collected some evidence that was produced against us in court later on.

Greathouse, Rubery, Libby and myself went on deck

and surrendered. We admitted nothing, contenting ourselves with saying that we alone were responsible for the ship and everything on board. They did not show the least surprise as they searched the ship and opened boxes containing our "knocked down" cannon and stands of firearms. They saw vast quantities of powder, shells, and ammunition of all kinds exposed with as much indifference as if they held a copy of the ship's manifest, which, in fact, they did have in their possession, through the treachery of Law. If anything further were needed to complete the knowledge that he had betrayed us, it was furnished by an unguarded remark of Captain Lees.

Our twenty fighting men, very much down on their luck, were found in a forward compartment. On our solemn declaration that they were employed only for service in Mexico none were prosecuted, and finally all were discharged with a "look out in the future" admonition from the officer in charge.

Some effort was made to sweat the four of us. We were cordially invited to step up like men and make a clean breast. All these courtesies were politely declined. We only asked to be advised what we were charged with, and the answer was sufficiently illuminating: "Why, piracy, of course." We were rather carelessly searched, so far as our persons were concerned. I was allowed to retain a small penknife, but one rather important thing was overlooked. In those days everyone carried a derringer, which looked like a sort of toy pistol, but was really one of the most deadly close-range emergency weapons ever invented by the evil genius of man. Each person had a pet place for keeping his derringer secreted, but handy. For myself, I carried one in a specially prepared pocket inside of the right cuff of my coat. Just a practiced twitch, and I could have it in my hand ready for use in an instant. This, as I said, in some way escaped the notice of my searchers, so though I was a prisoner, I remained fairly well armed.

All day long the wires around the world were telling of the great *Chapman* piracy project, happily nipped in the bud by the efficiency of Uncle Sam's government. One of the facts that gave it a peculiar interest was because John Bright's nephew was a participant. The story did not lose anything by age or travel. I had once a book of newspaper clippings relating to the *Chapman* affair, and a dispassionate reading of the more lurid descriptions would have satisfied anyone that Greathouse, Rubery, and myself were the most bloodthirsty pirates who ever cut a throat or scuttled a ship.

We were taken to Alcatraz and later to the old Broadway Jail. Greathouse was released after a few days of confinement on bail furnished by his relative, Mr. Lloyd Tevis. Among the pleasant incidents of our confinement were visits from Lieutenant Tompkins and Quartermaster Judson. Our late enemies became our best friends, brought us all kinds of necessaries and refreshments, including newspapers, periodicals, and books, and in every way sought to cheer us up and make our confinement less burdensome. Rubery, for his part, returned to Lieutenant Tompkins his letter of retraction, which the latter seemed very glad to receive, for in those days no man of honor cared to have documents of that kind floating around loose. Such incidents of good-will between men engaged on opposing sides in the Civil War prove to my mind that there was no fundamental line of cleavage, no real antagonism, in fact, between the North and South, and if there had been some power to steady the masses, instead of lashing them to fury, there never would have been a war.

As for Law, he had actually gone with us in good faith up to a certain point, then had a case of cold feet. It occurred to his sordid mind that a handsome sum of money could be obtained from the government, without any risk at all, by betraying his associates. He made a cold-blooded, mercenary bargain with the authorities through which he realized a small fortune, disclosed all our plans, and our steps had actually been dogged by detectives for days.

But the first day at Alcatraz I nearly landed Law. I was locked in a lath and plaster room. I had not been there long before someone began tapping on the wall. After several repetitions, thinking it might be Rubery, I asked, "Who is there?" The acoustics were admirable. A voice replied, "That you, Harpending? This is Law. I am under arrest. I want to tell you all about the awful mishap that prevented me from being with you on the *Chapman* last night."

The voice of the wretch drove me to absolute madness. I knew he wanted to draw me into admissions, probably had two or three witnesses with him in the room. I simply thirsted for his blood. As before mentioned, the searchers on the *Chapman* had overlooked a small penknife and a derringer concealed on my person. My first impulse was to take a chance shot at him through the plaster, but I thought of something better instantly. With my penknife I easily bored an opening in the wall.

"Law," I said, "there is something I want you to hear very distinctly, and I don't want to speak loud. Put your ear to this hole I have made through the wall."

If he had ever put his ear to that hole he would certainly have heard something very distinctly, and much louder than I intimated. Also, perhaps, this story would not have been written. But if such a fellow can have a good angel, she was not napping that day. Law did not put his ear to the hole, and a few minutes later I heard the door close behind him as he left the room.

CHAPTER XI

Technicalities Fall before True and Perjured Testimony and Author Is Quickly Convicted of Treason

We Find Consolation in Lack of Proof until a Foolish Remark Causes Weakling to Turn Informer

As I said, Rubery, Libby, and myself were brought from Alcatraz to the Broadway Jail, while Greathouse was enlarged on bail. We remained there over six months, while the government was preparing for our trial.

At that time there was published in San Francisco a paper called the *American Flag*. It perished peacefully after the war ended, but while it lasted, outclassed every publication of the North in downright ferocity, not alone to the cause of the South, but to every person of Southern parentage. It demanded that we be tried on a charge of piracy—a capital offense. But the closest examination of the law proved that no such accusation was tenable. The final indictment was for high treason. That also used to be a capital crime, but such a multitude of treason charges were brought during the war that Congress stayed the hand of the executioner and made the offense punishable only by imprisonment and fine.

Even that charge might have come to naught. Against us was the accomplice, Law, whose unsupported evidence was not sufficient. The armament found on the *Chapman* might have been intended for a filibuster expedition against a Central American state. The false custom-house papers might be explained in the same way, also the secret preparations for leaving the port, for the United States government was bound to intercept any illicit expeditions against friendly powers. Some general literature of an inflammatory "secesh" character was found on us, but our natural inclinations were a matter of public knowledge in San Francisco. Finally, the scraps of torn paper collected on the *Chapman* by Captain Lees and pasted together, while incriminating, were not complete and hardly admissible in a court of justice. In other words, while there was an ocean of suspicion, the prosecution could offer very little proof. Our best friends knew that the indictment was true enough, but to maintain it according to the rules of evidence was another thing.

The needed testimony, however, was supplied through some senseless talk of Greathouse. I have always contended that a man's worst enemy is his mouth, and there never was a better illustration. Greathouse visited us one day at the Broadway Jail. He was handsomely caparisoned, full of spirits, and I think had just risen from a good dinner, or rather lunch. Libby asked him anxiously about our prospects. "Well," said Greathouse, "they are not exactly flattering. I guess all of us will have to go to prison for a long term, but," he added somewhat grandly, "I will be able to buy my way out." He didn't say a word about the rest of us.

This remark started Libby to thinking. He was scared stiff before. Now he became a nervous wreck. He knew that Greathouse was powerful enough to be at large on bail. He knew that Rubery and I had influential connections. He was himself a poor fellow from Canada, adrift on the Pacific Coast, without a cent or friend. He saw himself made what we moderns call the "goat" for the whole *Chapman* incident and concluded that the wisest thing was to look out for his own hide. Somehow I have never had it in my heart to blame Libby overmuch for whatever happened. My impression is that he intended to "sit tight" until he thought himself left in the lurch.

Be that as it may, the day after the visit of Greathouse, Libby sent for the United States District Attorney, made a complete statement of all he knew concerning the outfitting of the *Chapman* and our designs against the commerce of the coast, adding, I am sorry to say, some details that were false.

This confession, brought on as I believe by the foolish talk of Greathouse, absolutely sealed our doom.

We were brought to trial on October 2 in the United States Circuit Court, Judge Stephen J. Field and Judge Ogden Hoffman sitting in bank, with an array of eminent counsel on each side. It did not take long to pick a jury in those days. The very dogs of San Francisco knew of the *Chapman* case, yet the twelve good men and true who swore they were unbiased were impaneled in less than an hour. Some of them were later noted. Here are the names: John Wheeler, Jacob Schrieber, A. S. Iredale, Samuel Millbury, Joseph D. Pearson, Joseph A. Conboie, G. W. Chesley, J. K. Osgood, James W. Towne, and W. P. C. Stebbins.

The evidence against us was overwhelming. Law and Libby told their stories in great detail. About half of it was rank perjury. They related conversations that never took place. Also incidents that existed only in their imaginations. Everything was set forth in its blackest light. The witnesses were well drilled and were not shaken by cross-examination. All of the other incidents were proved, the purchase of the ship through a custom-house broker named Bunker, the purchase of cannon and arms, the false manifest of the vessel, and the assemblage of a considerable fighting force. The government also proved that Greathouse and myself were citizens of the United States, not of the revolted states, while Rubery was classed as a common foreign adventurer. This, it seems, was necessary to establish the charge of high treason.

Our lawyers made the best of a bad job. They argued manfully many points of law, concerning which I have no recollection, except that they contended that the mere loading of a ship with arms did not constitute a crime any more than buying a pistol constituted murder; that in order to constitute the overt act the ship must sail for its destination. On this point the court held that leaving the wharf and laying to in the stream constituted "sailing."

Finally our counsel made a grandstand bluff. They declared that witnesses, then in Mexico, could clear up the whole transaction, but in the absence of these they were compelled to submit the case without testimony. None of us took the stand.

The lawyers unlimbered the usual forensic lore, illumined by bursts of fiery eloquence. Both the judges charged dead against us. However, Judge Hoffman threw me the following judicial bouquet:

> For the accused I feel a deep regret, especially for one of them who appears to have been animated more by a zeal for the cause which he has unhappily espoused than by the sordid and unworthy motive of enriching himself by the plunder of his fellow citizens. It is to be regretted that the courage and willingness to sacrifice himself for the benefit of his associates, a slight glimpse of which has been revealed by the evidence, have been wasted on an enterprise which is indefensible in morals as it is criminal in law.

It took the jury just four minutes to bring in a verdict of guilty of high treason.

A few days later we were brought into court and sentenced, each to ten years' imprisonment and to pay a fine of $10,000. The county jail was named as the place of our confinement until the government of the United States directed our imprisonment elsewhere.

As for Law and Libby, they were secretly placed on board a ship bound for China by the United States authorities, and were never heard of afterward, though I took some pains to learn their fate.

So ended the famous story of the so-called "*Chapman* piracy." I have given the details at some length because, while in itself rather trivial, it has been made to cut quite a figure in history. The facts have been so outrageously distorted that I thought it best for someone having full personal knowledge of every detail to tell the truth.

Libby's first name was Lorenzo. People often ask, "What's in a name?" Perhaps nothing, but I think otherwise. Lorenzo Libby helped to land me in prison. Lorenzo Smith did me up in a business deal, and I have unpleasant recollections of Lorenzo Sawyer, once on the federal bench of San Francisco. I never see a man christened "Lorenzo" without an impression that he will bear a heap of watching.

CHAPTER XII

Arrest of Accomplice Alarms Author and on Advice of Friends He Takes Flight

Amnesty Act Unlocks Prison Doors of Conspirators, But Fails to Bring Security

In war times, the American eagle was not a bloodthirsty bird. We began to have sympathizers, even among prominent Union men.

Greathouse was released after a brief confinement under a general amnesty act and upon taking the oath of allegiance. Rubery, a foreigner, could not take advantage of the amnesty act. However, at the request of John Bright, President Lincoln granted him a free pardon. But the astute statesman arranged that his precious nephew should not be involved in future trouble because of his Southern proclivities. He was placed on a Pacific Mail steamer and transferred at the Isthmus to a British ship bound for England. We had an affectionate parting, with the hope that we might again meet, a wish that was realized in a dramatic manner.

I alone was held, because it had been shown that I had a commission in the Confederate Navy. In almost exactly four months after my sentence, I was brought before Judge Hoffman and ordered released, under the same general amnesty act. The fine was likewise remitted. I am not versed in legal technicalities, but it seemed to me that the learned jurist stretched the strict letter of the law a bit in my behalf. Be that as it may, I always held the name of Hoffman in high esteem.

I was free at last, but only to enter into a new kind of bondage. I was broke. The fortune I had won by incredible good luck had vanished absolutely. What was worse, my mine in Mexico was abandoned during the French invasion and my title to it finally lost. It yielded wealth to other owners later on. I never saw my old chum Don Miguel Paredis again, but he kept his money and cut quite a figure in Mexican affairs.

When I stepped out of Broadway Jail, I was outwardly chesty but inwardly depressed, for I had just $8.50 to my name. Having been a free spender at one of the leading hotels in my capitalistic days, I went to the proprietor and frankly made a clean breast of my impecuniosity. He was overjoyed to receive me as his guest, gave me a fine room, and settled all my anxiety as to lodging and three square meals a day. That was nothing out of the common in the old days. But I would like to see the photograph of a man with nerve enough to make such a proposition to the manager of one of our first-class hotels in the present generation.

Still, my financial affairs gave me no little concern. I thought of writing to my father for temporary assistance, but there was an impediment even there. While I was raising cain and wasting my substance for the South in California, my progenitor was one of the strongest Union men in Kentucky. Some rather crisp correspondence had passed between us on that subject. Doubtless he would have been glad enough to assist or welcome the prodigal. But I was too proud to seek his aid.

This may justify a word of explanation. My father was a lineal descendant of Baron Harpending, who came to New York, New Amsterdam, with the original settlers from Holland. It was one of his ancestors who gave a lease for ninety-nine years to the Dutch Reformed Church of a piece of property in the business center of New York, now worth, approximately, three hundred million dollars. It was another Trinity Church case, with this exception, that there was no doubt about the lawful heirs when the lease terminated. My father brought suit to recover the property. That was one of the great lawsuits of the last century. Henry Clay, Daniel Webster, and Judge Underwood were my father's counsel. He won the case in the lower courts, but was vanquished in the Supreme Court of the United States on a technicality by a four to three decision. The family lost the vast property, but the church still displays the Harpending arms, as required by the lease of my ancestor, executed nearly 175 years ago.

My mother, on the other hand, was of the Clark family of Virginia, which settled in Kentucky over a hundred years ago. She was of the typical Southern strain. Thus, while my father, with his Northern antecedents, was an ardent supporter of the Union cause, I had the maternal blood in my veins. How we came to take opposite sides in the civil struggle was an instance of plain heredity, nothing more.

But that has nothing to do with this story. While I was worrying over finances, not knowing which way to turn and mighty downcast and blue-deviled, I was informed that my companion, Mr. Ridgley Greathouse, had been rearrested and was in custody. Ignorant of the charge and not having the wherewithal to fly, to say nothing of inclination, I determined to put on a bold front, walked down to the United States marshal's office, and asked him if he wanted me.

The mental processes of that functionary were of a leisurely nature. He looked me over with great care, scratched his head with a pen in a meditative way, slew a distant fly with a well-directed squirt of tobacco juice, and answered, weighing each word, "Well, not today, but I guess I will tomorrow. Where do you live?"

I gave him the name of my hotel and the number of

my room, which data being duly noted, we bade each other good day. To tell the truth, I was badly rattled. Nothing seemed more certain than that I was doomed to incarceration on a new charge, and, having enjoyed the public hospitality for almost a year, I had no stomach for any more.

Just as I reached my hotel, filled with these disturbing thoughts, a friend asked me to go with him to a business office. I sat down in the reception room, while my friend disappeared in a rear office. I could hear the sound of voices and the clink of gold. Presently my friend reappeared, carrying a small coin sack. "Harpending," he said, "you are certain to be arrested on some sort of an accusation. The Yankees will never let you stay at large. There are fifteen of us who have subscribed a hundred dollars each. Here is the money. We will also provide you a good horse and necessary equipment. You must leave tonight."

I asked the privilege of meeting my friends and was accorded the privilege. Several were not overburdened financially, and $1,500 was more than I could reasonably need. I selected the three richest, gave each of them my promissory note for $100, and returned the other contributions. We talked more or less of plans. I was advised to ride south to the neighborhood of Santa Cruz, across the mountains to the plains, and journey thence through the San Joaquin Valley to what was then known generally as the "Tulares," where such inhabitants as there were came mostly from the South and where, more important still, the law writs did not run.

So, shortly after dark, I made my preparations and proceeded to put as many miles between myself and San Francisco as the means of travel would permit. I turned down the horseback proposition, slipped on a southbound train in the evening, and before midnight reached its terminus, San Jose.

CHAPTER XIII

Hits for the Hills in Effort to Lose Pursuers, Passes One Good Thing and Stumbles into a Bonanza

Company of Soldiers Goes to Arrest Him; Is Taken into Camp and Very Soon After Everything Is Fine

I STAYED OVERNIGHT at San Jose at the house of a friend, a stanch Southern sympathizer, who had been advised by wire that he might expect a guest by the late train. The next morning, bright and early, I left with a companion and a stout team for Santa Cruz. On the outskirts of that town—pardon me, city—my companion left me late in the afternoon, directing me to a house of accommodation kept by a man I knew, of strong "secesh" proclivities.

I passed into the waiting room, where a number of men were standing. The proprietor received me with evident agitation and invited me to a room upstairs. "Mr. Harpending," he said, "the sheriff has received a telegram from the United States marshal to detain you if you pass this way. He will hear of the arrival of a stranger answering your description, as a number saw you enter my house. But"—and here he ripped out an awful oath, none of your feeble modern profanity—"I will send for some of the boys, and we will have one devil of a fight before he takes you."

I could see that the man was capable of anything desperate—I excused myself for a moment to get my luggage, slipped downstairs to the waiting room, took the small handbag that contained my personal belongings, went out the rear door and took the road toward Gilroy on foot. I hadn't any plan in view—just walked on, well into the night, until I was exhausted with fatigue and lack of food.

"Youth will not be denied" is an old saying. I passed a house where the lights were still burning and determined to seek a place of shelter. I knocked at the door. To my astonishment and joy, it was opened by a man called Clark, of Southern birth, whom I had met several times in San Francisco.

Clark received me like a long-lost brother, roused the household, had an old-fashioned Southern meal prepared that made me think of home, and an hour later I was sound asleep in a comfortable bed, safe among friends.

The next evening Mr. Clark accompanied me to Gilroy, where I was concealed in the hotel of a mutual friend for two days, waiting for a southbound stage that journeyed across the mountains to the San Joaquin Valley and thence to Visalia, in Tulare County.

There were two passengers on the stage when I boarded it, a gentleman called Byington and his friend, Thomas Staples. Byington's son was afterward district attorney of San Francisco for several years. The gentleman recognized me at once, and as we traveled along, I found that he wasn't a half-bad secessionist himself. He told me that he and his friend were bound to inspect a mine in which they were interested at a place called Kernville, about 125 miles southeast of Visalia. He advised me that it was a notable locality to "hole up" and avoid observation indefinitely; that the "ville," in fact, comprised only a few shacks, appurtenant to the mine, which was just in the early stages of development.

It was in the winter of 1864—the winter of the awful drought when scarcely a drop of rain fell in California. The weather was like midsummer. The great valley then only had a few straggling settlements. The vast prospect was unbroken, save when here and there a miniature whirlwind in the distance raised a spiral of sand skyward from the parched ground, or where a band of dust-laden, half-famished sheep, staggered on toward the mountains to escape a universal desolation. What a different prospect now. To one who saw those unbroken solitudes that are today among the busiest haunts of men, with fine cities, railroads, power lines, immense systems of irrigation, intensive agriculture, oil fields—everything, in short, that goes to make prosperity and a high civilization—nothing is more impressive of what a few brief decades of enterprise can bring forth.

As there was a small military post at Visalia, when we neared that town I made a detour on foot and joined Messrs. Byington and Staples to the eastward. We reached Kern without any noteworthy incident. The place was exactly as Mr. Byington described it—a collection of slab shacks to shelter a few men engaged on development work on the mine. This was known as the "Big Blue." It was an immense ledge of bluish quartz and was enjoying a boom on the San Francisco stock market. Byington was a type of the Californians of the sixties, who were ready to go into any mining stock venture, almost to the extent of their fortunes, without knowing anything more about the actual business than so many cottontail rabbits. At his request I examined the "Big Blue" and didn't like the looks of things at all. The superintendent raved about the richness of the ore, showed us fabulous assays, but there, staring us in the face, was a stamp mill that hadn't turned a wheel for months. I satisfied myself that while there were here and there small bunches of ore, sufficient to furnish seductive-looking assays, the general vein matter was far too low to be worked to a profit. As for the outlook, that was another thing. The mine might prove to be rich at a greater depth, but the chances were at least 50 to 1 that it wouldn't. I advised Byington to unload his stock while he could, which he did to his great advantage. A few months later, "Big Blue" stock certificates weren't worth picking up in the street. Nevertheless, "Big Blue" was the inspiration for several later mining-camp crazes. Among others, Senator J. P. Jones of Nevada dropped a good-sized fortune in it.

I became known at Kernville, and as people were traveling to and fro, it was certain that my retreat would soon be common property to my enemies as well as my friends; so I decided to seek solitude and efface myself. With three companions of a roving nature, we struck out for the mountains, and for some time enjoyed the delightful, carefree Bohemian existence that cannot be found in many places outside of California.

I hadn't forgotten my old mining habits. One day I picked up a number of fragments of quartz, broken by the weather from a ledge that had a likely look. I took these to our camp, crushed them in a primitive way, "panned" the product, and stood aghast with astonishment at the result. A long, heavy "tail" of gold in the pan told that the rock must be worth hundreds of dollars a ton.

There is something about gold—just the metal—that makes people forget about everything else in life. A little prospecting showed us that we were in the heart of a great gold-bearing district, with surface croppings of such value that all a man needed for working capital was a pick, a pan, a couple of hammers, and a mule to carry the rock to water, where it could be hand-crushed and washed. Even the mule could be dispensed with if one did not mind the labor of shouldering an ore sack for a short distance every day. We were rich and gold mad.

Realizing the importance of the discovery, I sent one of my companions to collect enough men to form a mining district under the existing laws. These assembled, we perfected an organization and elected officers. Somewhere in the Book of Genesis, mention is made of a river in Paradise running through the land of "Havilah, a country rich in gold." We were shy on the river, but the balance of the quotation seemed appropriate enough, so I christened the proposed town "Havilah." The district was called "Clear Creek," under which title it was famous for many a year. Also, I showed judgment and forethought in a real-estate way, claiming and staking off a natural townsite.

I didn't dare to go down into the settled district to purchase anything like machinery for fear of arrest, but we constructed rude arrastras, primitive Spanish quartz mills, and began to turn out gold bullion in astonishing amounts. Something concerning a new gold discovery began to leak out, and occasional prospectors joined our camp. We had no end of provisions, plenty of fresh meat, and sort of kept open house. All in all, it was about the best-ordered mining camp I ever saw.

But the big boom for the camp came through my old journalistic enemy, the *American Flag*. Word came to it somehow that I was located in the mountains back of Kern City, ostensibly engaged in mining. Straightway it gave me a terrific blast, claiming that mining was only a cloak for a new piece of deviltry I was hatching. The effect of this was that it located me for a lot of my Southern friends who really believed that I was organizing a band to fight through to Texas, and as a consequence they began to swarm into Havilah in large numbers. Nearly all my fighting men of the *Chapman* were among the first arrivals. Also several Northern men, fired by the word "gold," took a chance of entering into an alleged stronghold of conspirators. They would have marched into hell, just the same, for gold. They had the same reception as anyone else. Far up in the mountains, away from strife and faction, these men mingled in perfect amity and good fellowship. It was another illustration of what I said before—that if the people had been left to settle matters in their own way, there never would have been a Civil War. Chattel slavery in the South was fast dying at the root. Another decade or so would have seen its finish. And the real question of slavery was not settled at all. There have grown up other forms of slavery far more odious and soul destroying than the mild system maintained, with very few exceptions, in the South. It was the agitators and demagogues on both sides, who never fought at all, upon whom must rest the responsibilities of our war, just the same as in nearly every historic struggle. Strangely enough, these men are commonly canonized instead of being held up to the execration of mankind.

Havilah was fast becoming a large proposition. Its trade was eagerly sought for, and pack trains of supplies were arriving daily. But the more it grew, the louder and longer raved the *American Flag* about the band of outlaws in the mountains, headed by the piratical Harpending. So specific were the denunciations that at length, they seriously attracted the notice of the government. Finally a detachment of troops stationed at Visalia was dispatched to drive us out.

We had timely notice of this kindly intention. I had been recognized as a sort of leader, partly because of my position as the largest owner of the district, partly because of the newspaper notoriety, which had given me the character of a daring adventurer—the character that impresses the rough natures of a mining camp. All the miners were called together. Lookouts were stationed down the canyon to give notice of the approach of a hostile force. I had decided to adopt Albert Sidney Johnston's strategy and try the moral effect of a surprise.

But the wily soldier in command did not come by the usual route. Early in the morning we heard the sound of cavalry tramping down the mountainside. We were prepared for that. The officer and his troopers, about eighty in all, walked into an ambuscade and suddenly found themselves confronted by four times their number, raw-boned, bearded, athletic miners, each armed to the teeth. I stepped forward, saluted the officer, who seemed a trifle rattled, congratulated him on being just in time for breakfast, and carelessly asked him if he had lost his way.

The officer replied, in a surly fashion, that his business was to disperse a band of cutthroats and rebels. I answered that he could hardly mean us; that we were peacefully pursuing a lawful occupation; that we were ready to submit to legal authority, but must first know the nature of our offense. I urged him to examine the camp, interview some well-known Union men who were with us, and satisfy himself that we were neither outlaws nor rebels. There was nothing for him to do but make the best of a bad bargain. The troopers rode to camp with their miner escort, had a jolly good breakfast, with more or less joshing on either side, and that part of the incident closed in a happy way. But the officer declined to be comforted. He was clearly mortified at our successful strategy.

We all knew that this was only a respite—that more serious trouble was ahead. But, in fact, it proved the camp's salvation. A gentleman called Sumner—I forget his other name—a Northerner of character and standing, who knew all about our case, came to our defense in San Francisco. I also sent a full statement of our case to my friend, Colonel Crockett, later a justice of the Supreme Court. These two waited on General McDowell, in command of the Department of the Pacific, and so far convinced him that he sent rather peremptory orders to Visalia not to interfere with us further, except on direct command.

At the same time, I was advised that I was free to go to and from San Francisco; that there never had been, in fact, a charge against me; that the rearrest of Mr. Greathouse was in no way connected with myself; that the United States marshal had only been advised to keep an eye on me; that he had only wired the sheriff at Santa Cruz to do the same.

In other words, I had fled from a man of straw—from a lighted pumpkin head in a dark room—and had stumbled over a fortune.

With sufficient money I made haste to San Francisco, paid my trifling debts, not overlooking the hotelkeeper, bought a quartz mill and appurtenances, rushed it down the valley, and had the stamps falling in record-breaking time.

The year 1865 was a busy one for Havilah and the Clear Creek mining district. It became a heavy gold producer—miners, capitalists, speculators swarmed into it from all over the Pacific slope. I laid out my townsite in due season and sold it out at fancy figures. The main street brought an average of $20 per foot. A boom was on all along the line. I was offered fancy prices for my mining claims. I let them go. My principle was to avoid what is vulgarly known as "hoggishness." When I could make a million by a business turn, I considered it a good day's work.

But, as a matter of fact, I only cleaned up with $800,000. That is what I banked in San Francisco long before the end of 1865.

The town of Havilah prospered mightily. At one time it must have numbered nearly 3,000 inhabitants. It was a brisk center, with hotels, livery stables, large merchandise stores, lawyers, doctors, preachers, open gambling houses, hurdy-gurdies, saloons, banks, bagnios, and the other evidences of advanced civilization.

Not only that, but its enterprising inhabitants appeared before the next legislature and asked for the creation of a new county. Though by that time a permanent resident of San Francisco, I assisted in the passage of a bill that cut off from Tulare the county of Kern and named Havilah the county seat. It so remained until the decline of mining and the growth of agriculture in the lowlands moved the capital to Bakersfield.

These statements can be verified by official records of Kern county and of the town of Havilah, which, I presume, still exist. Also by the testimony of many people still living.

Such were the tricks Dame Fortune played me in a period of a little longer than a year.

I was literally chased from absolute poverty into the possession of nearly a million dollars.

I discovered a great mining district and founded a thriving town.

And if the matter of paternity is ever brought up in court, it will probably be proved to the satisfaction of a jury that I am the father of Kern County.

CHAPTER XIV

Decade Between 1860 and 1870, Next to the Gold Age, One of the Most Stirring Times in History of State

Realization Had Come that Mineral Riches Formed Smallest Part of Resources; Outlook Was Bright

Late in the summer of 1865, I took up my residence in San Francisco. The war was over, the country settling down after the intoxication of a terrific struggle. But one fever was only followed by another, so far as I was concerned. I was barely twenty-five, but far older than my years. In fact, I never had any youth at all. From the time when I ran away from college to join Walker's expedition against Nicaragua, I was called on to meet problems that required a man's decision, and so became one, long ahead of time. But I was brimful of a restless ambition to make my mark—to become one of the great central figures in working out the destiny of the Pacific Coast.

Those were stirring times indeed. Few seem to understand that the decade between 1860 and 1870 was, next to the gold age of the fifties, the most important in the history of California. It was the period of transition from the fierce exploitation of the pioneers, who looked only on the region as a thing to be despoiled of its treasures and to be abandoned. It saw the silent valleys changed to broad oceans of waving grain. It saw the foothills crowned with thrifty vineyards, saw the sure foundations laid of a great fruit industry, saw the beginning of systematic irrigation. It saw the port of San Francisco crowded with masts of vessels to carry its new-found wealth to distant lands, saw a mighty foreign commerce develop, saw the treasures of the Comstock Lode unlocked, saw a railroad stretch from the Atlantic to the Pacific. And men arose to meet the new conditions. A splendid line of merchants seized the opportunities of trade. Isaac Friedlander opened the markets of England for our wheat. Macondray Brothers built up great business interests in the Orient. The trademark of William T. Coleman & Company was a guaranty of their goods throughout the civilized world. These names are only typical of many. A new race of mighty miners developed, men like George Hearst, J. B. Haggin, Lloyd Tevis, Alvinza Hayward, G. W. Grayson and others, whose activities extended to Utah, Nevada, Arizona, Montana, and to distant Mexico, pouring a fresh river of gold and silver into California.

The drift of population of the growing city also changed, and the westward movement began, which will only be bounded by the ocean.

In short, nearly all we have in northern California today in the way of industries and enterprise can trace the starting point to that age.

It was an intense, booming, hopeful decade, a period of great events and great men, when everyone at last realized that gold was the smallest part of the state's resources and the outlook as broad as the horizon of mid-ocean. I do not wish to interrupt the narrative to dip into general history, but it may interest the reader to have a glimpse, as we jog along, of real things and the live people of what I may be pardoned for calling the old but recent times.

All old Californians can recollect the now-faded glory of Montgomery Street. Stretching barely from the foot of Telegraph Hill at Jackson Street, nine blocks, to a full stop at Market Street, it was really the whole town. During the busy hours of the day you could meet there every man worth knowing in San Francisco and in the afternoon every woman with a pretty face or a handsome gown to show. This gave a wonderful facility for acquaintance and general good fellowship. Everybody knew everybody. That was what made the old San Francisco the most charming and fascinating city in the world from a social standpoint. It was not alone the most brilliant society I ever encountered in an experience that has covered most of the world, but there was a freedom and heartiness in general intercourse that could only be explained by the conditions under which people lived.

Into those nine blocks and, to a less extent, into one block on either side of some of the intersecting streets, like California, Pine and Bush, a vast business was huddled, no less remarkable for its vast extent than for its cosmopolitan, or rather heterogeneous character. Banks, commercial houses, stock exchanges, brokers' offices, courts, public buildings, the leading hotels, retail stores, public libraries, theaters, music halls, the two great social clubs, nearly all the lawyers in town, the leading doctors, and probably the finest saloons in the world were mixed up inextricably like a huge human menagerie broke loose. Not to be on Montgomery Street or within half a block of it was to be classed as a business, professional, or social pariah.

Of course real estate values soared skyward. It was hard to estimate what Montgomery Street frontage was really worth, but there were actual transactions as high as $6,000 a front foot, nearly as high as the present selling price of choice realty on Market Street. Rents likewise were enormous. Considering how little the landlord gave in the way of conveniences to his tenants, these rents were much higher than they are today.

To relieve this tremendous congestion was one of the

problems of San Francisco in the sixties. No one had the perspective to forecast cable and trolley cars climbing all kinds of grades and peopling the hills with homes. All we could see was an extension south, and for that purpose the city was badly laid out.

The battle cry in the early sixties was "Montgomery Street straight." The all but universal wish was to run the great street, broadened to a wide avenue, in a direct line to Connecticut Street, far to the south. Tremendous efforts were made to carry through this project in a peaceful way. Several times it was near accomplishment, but just as often fell through, owing to some recalcitrant property owner. The main obstacle was the large block of land on Market Street where the Palace Hotel now stands. This was owned by the Catholic church and had been reserved for the construction thereon of a great religious edifice.

As soon as I got my bearings in San Francisco, I saw at once what a vital question was involved and what a grand opportunity was there to win not alone fortune, but fame. I carefully surveyed the situation from every standpoint and finally hit upon a scheme which would carry out the original design of "Montgomery Street straight" and avoid the opposition hitherto evolved. As what follows forms one of the interesting bits of San Francisco's history, hitherto untold, and the city's present status was greatly influenced by my plans, which, however, were only carried out in part, I will give an outline of one of the largest real-estate transactions of a far-reaching character ever conceived and partly completed in the history of San Francisco.

CHAPTER XV

First Speculator to Figure that Market Street Had Future; Buys Several Choice Lots for a Pittance

Earthquake Plays Important Part in Big Deal; Timid Citizen Sells Out in Hurry and Loses $350,000

In the early sixties no one thought of Market Street except as a disfigurement to the city and a broad impediment to its progress. It began almost nowhere, at an unfrequented section of the water front, where the dullness was relieved only by the arrival and departure four or five times a day of a ferryboat owned by Charles Minturn, which transported a few straggling passengers between San Francisco and the small village of Oakland across the bay. If I recollect right, the fare was fifty cents each way. It terminated—so far as traffic and settlement were concerned—exactly nowhere, in the desolate sand hills beyond where the Flood Building now stands. The "gore" streets, like Post, Geary, and O'Farrell, that now pour a human tide into the city's big artery, were settled only for a few blocks westward; beyond that there was solitude. The roadway of Market Street was an abomination even in the old days before "good roads" became a slogan. The sidewalks, if any, were wooden, and mighty poor at that.

I was about the first real-estate investor, to which business I turned my attention, who realized in a sort of vague way that Market Street had a future. I had a shrewd idea that the insistent plan for "Montgomery straight," on which so many based their hopes, was doomed to disappointment, but as I said in the last chapter, I saw a way out of the dilemma. With this in view, I began to pick up Market Street frontage from First Street west till I owned 800 feet. I also bought the Sutter Street gore, where the skyscraper has since gone up, at a public auction, paying $86,000 for the same in hard cash, which was considered a top price.

Meanwhile, I was quietly buying a solid block of land straight through from Howard Street to Market. It was broad enough to allow a wide street to be laid out in a line directly opposite to the ending of Montgomery and thus change so many back yards into frontages on a fine thoroughfare—an extension of the city's crowded mart. I wasn't a prophet or the son of a prophet, but the enterprise looked good. Not only that, but it seemed certain to me that the extension of Montgomery Street, once begun, either "straight" or at an angle, would be pushed ahead by the force of public opinion to its proper terminal, the water front of the bay.

Most of this immense property, outside of the Sutter Street gore, was gathered in at prices so pitifully small that they would test the credulity of the reader. Excepting one other piece of property, the total investment was less than $500,000, and the wretched buildings on it yielded a net income of 1½ per cent per month on this amount. However, this was looked on as a very poor return in those days, when 2½ per cent per month, compounded with the regularity of fate, could be obtained on well-secured loans. The money market was considered easy when borrowers could be accommodated on those terms. It all seems like some Monte Cristo story, but let me tell the young, ambitious reader that there are just as splendid opportunities staring him in the face today, just waiting to be taken into camp. The region around the Bay of San Francisco is destined, beyond a doubt, to become one of the greatest world centers of population, commerce, business, and production. The possibilities have barely been touched. Take it as the judgment of a close observer, with wide experience, that now is the time to get on board—using, of course, sound common sense and intelligent foresight. If I were twenty-five instead of seventy-six, I would like to give the public an object lesson of how to make money.

But I was quite a while securing the Market Street frontage necessary to carry out my plans. How I finally succeeded is quite an incident, well worth recalling, although it is ahead of my story.

On the south side of Market Street, exactly opposite the terminus of Montgomery Street, stood a vacant, fifty-vara lot, owned by a well-known old pioneer by the name of Selim Woodworth. I had bought sixty feet adjoining the church property where the Palace Hotel now corners. The Selim Woodworth lot was next to that. Now the owner had a very hard head and a rather top-heavy idea of the value of anything he possessed. Off and on I was negotiating with Woodworth for more than a year and a half, sometimes personally, sometimes through a broker, but always ran abruptly against the same proposition. "If you want that property, it will cost you precisely half a million. There is nothing further to be said on the subject. Let's turn the conversation to something else." Nobody could have been more courteous or more firm.

I had been bumping up against this brick wall so often and was so helpless to carry out my plans without the land in question that I was on the very point of paying what was then a price out of all reason or proportion to existing values when something happened to change Mr. Woodworth's estimates in a way quite novel and picturesque from a business standpoint.

The earthquake of 1868 wasn't much alongside of its successor of 1906; nevertheless, it was quite a jolt. Some rattletrap buildings collapsed, many others were cracked from roof to foundation, an immense number of chimneys were overthrown, and a few people killed. The most disquieting feature was that the earthquake didn't know when to stop. There was the first big damaging shock, but after that, every ten minutes the earth gave a jolt, quite hard enough to send multitudes scurrying into the streets. There was not an interval long enough to allow the average man to gather his wits. After a night of agony and suspense, most people were weak as kittens and speechless.

I went downtown in the morning after the quake from my home on Rincon Hill. I really was not much disturbed after the first lurch, for I had read somewhere that the minor tremors succeeding the initial shock never need be apprehended. One of the first men I met in the business section was Selim Woodworth. He was carrying a handbag, and his face showed evidence of mental strain. I asked him where he was going.

"Where am I going?" said Woodworth. "What a question to ask! Why, I am getting out of here before the earth swallows me up. When do you leave?"

I told him I didn't intend to leave at all, whereat a look of deep craft came over Woodworth's face. "Look here, Harpending," he said, "how about that Market Street lot? Do you still want to buy?"

I laughed as if in scorn. "Who on earth," I said, "would want to buy a lot that may be a hole in the ground by night, reaching through to China. Besides, you have always been so unreasonable that no human being could deal with you."

"Well, make me an offer anyway," he replied, "you will find me reasonable enough."

I pondered for a moment before I answered. "Well, I was thinking of offering you $150,000," I said, "but that's too much. Still, just to help you out in a neighborly way I might strain a point and give you—"

I never had time to finish the sentence. "Oh, for God's sake," yelled Woodworth, "don't screw me down at a time like this. Make it $150,000 and we'll close the bargain here."

We shook hands on the spot. Together we went to the Bank of California, where a formal contract for a deed to the property was drawn, and this, at Mr. Woodworth's request, was guaranteed by a high official at the bank. I deposited $150,000 in escrow. In due season the deed was sent on to Mr. Woodworth in Europe, was returned properly executed, and the famous fifty-vara lot passed into my hands.

This is the veritable story of how I acquired the frontage on Market Street which enabled me to open New Montgomery Street through my property to Howard. I was certain it would soon be extended to the bay and solve the problem of the sixties—"Montgomery South." At Market Street corners of the street I opened the Palace Hotel and the Merchants' Bank, two of the finest buildings in San Francisco, now stand.

I made New Montgomery Street, as it stands today, a free gift to the city of San Francisco, and it must remain a permanent record of my existence. I may add that I had to scatter numerous shekels among the "boys" before the gift was finally accepted.

Also, the earthquake literally shook $350,000 out of Selim Woodworth, which he would have received otherwise in a few days.

CHAPTER XVI

Montgomery South Deal Comes to Notice of Ralston, Who Buys Quarter Interest in Real Estate Project

Long before the events narrated in the last chapter a most important person became a character in this narrative. In one way or another I had become quite a figure in the business world of San Francisco; I took a flyer at several things in a speculative line, always made money at my ventures, and was generally looked on as what we now call "a comer."

But it was entirely because of my large and peculiar real-estate investments that I attracted the notice of the great central figure of California of that day—one who always wished to be associated with any of the large movements of his time.

Almost from the date when I first had money enough to make it inconvenient to carry it on my person I kept my account at the Pacific Bank, of which Governor Burnett was president. But I had watched the ascendant star of William C. Ralston as it put out of sight all the lesser luminaries. I had a young man's admiration for his dash, energy, and success, and I was pleased when I received a letter from the gentleman saying that he would like to see me, at my convenience, at the Bank of California.

We met by appointment. I had known Mr. Ralston before in a purely casual way. This was the first time we had touched in business. He had a swift, offhand fashion of saying pleasant things—not flatteries, but things that put a man in good humor with himself, and thus he spoke to me of his desire to be abreast with the active men of the city, to be able to aid and co-operate with them. Then he had a word to say about my real-estate ventures and in a very natural course led up to a general conversation on the subject.

Mr. Ralston's manner entirely won my confidence. Besides, we had a direct way of doing business then, quite different from the dark-lantern methods of today. I simply laid down my cards on the table, face up. I told Mr. Ralston exactly what I had in mind; that my purpose was to solve the great problem of "Montgomery South" in a new way; that I considered it a matter of vast importance for the city's future and one certain to bring fortune to the successful promoter.

Mr. Ralston listened with deep attention, with an occasional word or nod of approval. When I concluded, he leaned back in his chair in a meditative way and thought a minute. "It looks like a noble game," he said at length. "Now, how would you like me for a partner?"

I was just a bit astonished at the proposition, but I was gratified that the financial autocrat of the Pacific Coast wanted to climb into my band wagon. The arrangements were made with less "jockeying" than now takes place over a thousand-dollar transaction. I made a complete statement of my investments, which Mr. Ralston accepted and made me an offer based on cost, plus a very handsome profit, for a quarter-interest. I accepted also, in an offhand way, and deeds passed to correspond covering all my real estate involved in "Montgomery South." It was understood that we would stand together to push the new street through to the bay and that in this project I should have the practically unlimited support of the Bank of California. Our holdings were merged into a corporation known as the Montgomery Street Land Company.

Thus I became associated with this strange character, who has been dead almost forty years yet whose name is

WILLIAM C. RALSTON

President of the Bank of California, the first dupe in the diamond fraud

still a household word to thousands and bids fair to be remembered long after those who pose as the truly great are asleep in forgotten tombs.

I spoke of Ralston as a "strange" character. But the adjective doesn't fit the case at all. There was just one Ralston in California. Perhaps his counterpart never lived before. It would be far beyond my powers to draw a sketch of a man so many sided, but here and there some traits cropped out so prominent that they could scarcely miss the observation of a child.

Ralston had a marvelous head for business. The most difficult problems of finance were as simple to him as the alphabet, and his mind cut through all perplexities and obstructions straight to the truth. Had he possessed a few less red corpuscles in his blood—been a plain, downright financier, I am certain that he would have grown beyond the narrow environment of the Pacific Coast and become one of the world's money kings. But he had an odd supplement to the cold-blooded faculty of making money, a sort of richly Oriental imagination that looked far beyond the mere acquisition of a pile of cash.

For one thing, he had a passionate, almost pathetic love for California. He wanted to see his state and city great, prosperous, progressive, conspicuous throughout the world for enterprise and big things. I think it was this imagination, this ambition, that kept hurrying him into one big undertaking after another, many of which were way ahead of time. While he was stacking up money in one direction with the skill of a great native-born financier, it was leaking out in various other ways for rolling mills, vast hotels, watch factories, woolen mills, furniture factories, and what not. It was only an ambition to say that San Francisco had the grandest and largest hostelry in the world that had prompted him to build the Palace Hotel. And so on down the line. He tried to do everything and, like others, failed in the end.

With all of his tremendous business activities, I could never think of Ralston except as a big, overgrown boy. He had an elasticity and buoyancy of spirits very seldom seen beyond the teens and a youth's eagerness in the pursuit of pleasure. Nothing seemed to disturb his imperturbable good humor. He was at once the best winner or loser in the world—could pick up or drop a million with equal gaiety and nonchalance. He always smiled in conversation, but in moments of repose his features settled into an expression that was half-thoughtful, half-sad.

Where he shone most perhaps was as a "mixer." He had wonderful manners, frank, cordial, magnetic, and handed out the same quality to everyone alike. He avoided, either by design or inclination, all the pomposity and circumstance of greatness, even went out of his way to be extra gracious to those who seemed a trifle embarrassed in his presence. In this way he endeared himself to a small army of young men and to some of the still more youthful highbinders who used to visit the Bank of California and stand up the smiling financier for baseball-club uniforms and other all-important incidentals upon which the fame and glory of California hung. As to the industrial classes, they simply worshiped Ralston. He was their constant provider, philosopher, and friend. It seemed to me that he knew half of the working population by their first name, and he was known among them familiarly and affectionately as "Bill" Ralston. It is a sad commentary on human nature that in the hour of his misfortune the men he had enriched took to the tall timber. Only his humble friends proved true.

This only gives the faintest glimpse of Ralston. For years I was his intimate associate. He was a man of many friends and many business connections, but I think he gave his entire confidence to only two men, Maurice Dore and myself. In all our relations I always found him punctiliously honorable and truthful, and though the acquaintance was a costly one to me, I hold his memory in affectionate remembrance—as I did when I last saw him forty years ago.

And even admitting all that his traducers charge, after death had silenced his voice forever, I would still say that he deserved a statue in Golden Gate Park as the most effective friend the state of California ever had. But before I close this story, I hope to make it plain that a cruel wrong was done his memory and let the truth come out at last.

CHAPTER XVII

Sharon, Too, Becomes Associate of Famous Pioneer; This Chapter Tells How Great Panic Was Averted

Ralston Lays Foundation for Huge Fortune of D. O. Mills by Making Him a Bank President

RALSTON HAD two business associates—I might almost call them familiars—William Sharon and D. O. Mills. D. O. Mills was a man of some fortune, worth perhaps half a million dollars. He was about to leave for the East to settle down somewhere under his own vine and fig tree when Ralston took him up. The latter was just organizing the Bank of California, had no ambition for titular dignities, and offered Mills the place of president. He promised that the job would be a sinecure—that he would do all the work. Mills accepted. That was the foundation of his huge fortune. But he was a mighty cautious speculator in those days. He tried his hand at a number of ventures, sometime invested large sums, but always required a guaranty against loss. Strangely enough, this used to be given him often because of the conservatism associated with his name. That, however, did not last long. He became a bold, ambitious original operator on his own account. He had fine personal habits, but was just the opposite of Ralston—unemotional, cool headed, and austere.

Sharon was quite a different character. He was from the same part of the Northwest as J. D. Fry, an uncle of Mrs. Ralston, and so received a favorable introduction to the famous financier. Ralston sent Sharon to Virginia City during the early flush times as an agent for his mining interests, and when the Bank of California was organized and a branch located on the Comstock Lode, Sharon became manager, a position of great prominence and power. He was a daring, spectacular plunger, though a very shrewd one, made big money from the outset, and, with his general foreknowledge of conditions, really took no chance at all on great stock-market deals. With unexampled rapidity he accumulated a fortune of millions. Having no pet hobbies to interfere with accumulating money, before the sixties were over both Mills and Sharon probably possessed larger fortunes than their chief.

As I have said, Sharon's personality was very different from that of Mills in many ways. He had some habits that an anchorite might not approve. Among other things he was devoted to the great national pastime—draw poker. Many legends of his prowess, of his bewildering bluffs and high-class technique were long-fragrant memories of the Comstock Lode. It is related that a friend of Ralston with a moral turn warned him that his agent at Virginia was a notorious, abandoned, and dissolute poker player. The banker listened with absorbed interest. "Does he win or lose," he asked.

"My information," said the informer, "is that he almost always wins."

"Good," said Ralston, "he's the very man I want." The three were associated in many enterprises of the first magnitude. They had combined resources, speaking very conservatively, of not less than thirty-five millions, which is a big bunch of money even today. They formed an irresistible power in California until the railroad dynasty succeeded them. Still, I have reason to believe that Ralston never gave these associates his full confidence and that the relation became irksome, if not oppressive.

I judge the former from an unrecorded incident that

WILLIAM SHARON

A leading Comstock figure, former U. S. Senator

ought to be remembered in financial history—an incident that just prevented a crisis far more terrible than that of 1907.

It was in the year 1869. Ralston had loaned the railroad people some months before $3,000,000, with which they pushed their line to Ogden, Utah, adding a hundred miles to the Central Pacific in its cross-country race with the Union Pacific for mileage. This large sum had gone out of the state absolutely. Also, two millions had taken wing for South America to finance an investment there. Things were already a trifle tight when in July, 1869, Jay Gould's famous "gold corner" raised the yellow metal to a huge premium, and the gold coin of California was drained eastward, as through a sieve. The banks always carried a large amount of gold bars, but this was not available as coin, for the mint happened to be shut down, pending a change of administration.

The situation need not have been serious, for tucked away in the United States Subtreasury were $14,000,000 in gold coin. It seemed the most legitimate transaction in the world to deposit gold bars in the Treasury and carry away an equal value in coin. But President Grant, who was rather new on the job, for some unaccountable reason absolutely refused to sanction the transfer, although the bankers almost burned up the wires with their appeals. An uneasy feeling was over the town, the overcharged atmosphere of panic, apt to break loose at any moment into a resistless storm.

While the tension was at its height, I called at the Bank of California one afternoon and was ushered into the private office of Mr. Ralston. To tell the truth, I was feeling the pinch myself and wanted to know something of the outlook.

The banker said I was just the man he wanted to see.

"If things go on as they are," he said, "every bank will be closed by tomorrow afternoon. Not one of us can stand a half-day's run, and all will go down in a heap. Then look out for hell in general to break loose. This will happen if I don't get a million dollars in coin in the vaults tonight. But I intend to get it and want you and Maurice Dore to help. Be at the bank at one o'clock tonight, and put on an old suit of clothes, for you will have plenty of hard work to do."

Dore and myself met by appointment shortly after midnight. We were utterly mystified. Together we tramped through the deserted, dimly lighted streets. It seemed just like old times—the time we boarded the *Chapman* to become privateers. We found Ralston at the bank with one of its trusty officials, still alive and prominent in San Francisco. The financier was in high spirits, but counseled caution. We walked noiselessly to the United States Subtreasury, then located on Montgomery between Sacramento and California streets, where the Selby offices afterward stood. A dim light was burning within. Mr. Ralston asked us to halt a few paces from the entrance; then, to our great surprise, he opened the door of the Subtreasury, without challenge of any kind, and closed it after him as he stepped inside. Presently he emerged with several sacks of coin. "Take that to the bank," he said. "The gentleman there will give you something to bring back."

The party at the bank received the cash, tallied it, and handed us gold bars for the same value. These we took to the Subtreasury, where we found Mr. Ralston smilingly awaiting us with a new cargo of sacks on the sidewalk. We turned over the bars and made another journey to the bank.

Thus, at dead of night, passing to and fro, we transferred in actual weight between the Subtreasury and the bank nearly five tons of gold. We did not get quite as much as Ralston wanted before the light began to break. It was a heartbreaking job from a physical standpoint. I was young and athletic and stood my end of it in good shape. But Maurice Dore was of sedentary habit, soft as mush, and he was on the verge of collapse. He was nearly chest foundered and had a sway-back appearance for a month. During all this time not a person passed to interrupt us. This was doubtless due to a prearrangement with the policeman on the beat.

When the Bank of California opened the next morning, a rather ominous-looking crowd was in waiting. Lines began to form behind the paying tellers' windows. It wasn't a "run," but a "near-run." Ralston appeared on the scene and looked annoyed as he said, "Why are you making so many of our customers wait on a busy day? Put more tellers on the windows and have your coin on hand." More tellers went to the windows. Porters brought tray after tray from the vaults. It was amazing how the crowd changed their minds about wanting their money and melted away. And all over the troubled city the report spread that the Bank of California had coin to burn, and the news caused a general relief.

Nevertheless, a serious run started on one of the leading banks. Ralston hurried to the spot, mounted a dry-goods box, and addressed the crowd. He told them they were doing the bank and the city a great injustice. He declared that the bank was absolutely sound—which was the truth. He further told the crowd that they need not wait for a line-up. Just bring their books to the Bank of California, and they would be accommodated with the cash there. Again the crowd slunk away, abashed.

Thus a tremendous panic, the consequences of which might have been world wide, was averted by a bold front, a nervy bluff backed by a million in cash. Three days later President Grant reversed himself and allowed gold to be exchanged at the Subtreasury for cash, which settled all anxiety. This was brought about through the agency of Jesse Seligman, the New York banker, who gave the President a banquet and then showed him his mistake.

But neither Mills nor Sharon, who were leading officers of the bank, ever knew how Ralston gathered in nearly a million dollars after banking hours that day. All the satisfaction they ever got was that a kind friend had come to the bank's assistance.

CHAPTER XVIII

"Big Four" Intervenes and Sets Up Obstacles; Ralston Acts as Mediator and Is Badly Gold-bricked

Railroad Madness Results in the Narrator Securing Franchise for Line from Sausalito to Humboldt

Way back in 1868, the Legislature passed a bill giving a franchise to a corporation organized under the name of the San Francisco & Humboldt Bay Railroad Company to construct a railroad from an indefinite point on the Bay of San Francisco to Eureka, in Humboldt County. The franchise was coupled with a provision that the electors of the counties through which it passed should be authorized to vote a subsidy in bonds of $5,000 per mile, payable as every section of twenty-five miles was completed. That was about enough to pay for the rails. The franchise was later extended to the water front of Sausalito, but that was surrendered to the Sausalito Land and Ferry Company.

The franchise was held by Fred McCrellish of the Alta, J. F. McCauley, a well-known business "rustler," General Connor, a temporary sojourner from the Northwest, and I think H. T. Templeton had a small interest. None of them had any capital to speak of, and they had no other design than to peddle the franchise to someone who had.

Of course the promoters had done nothing in the way of construction, and the rights were in a fair way to lapse when Fred McCrellish drew my attention to this paper property and asked me to make an offer. The Central Pacific was then nearing completion. Like most people in the state, I was railroad mad, and being on the lookout

ASBURY HARPENDING

Taken during his active career in San Francisco

for everything good, I referred his proposition to an expert. The report of the engineer was very favorable, and when I found they wanted only $20,000 for all their rights and franchises for a railroad from Sausalito to Humboldt Bay, I readily closed the bargain and bought them out, all except one-tenth, which J. F. McCauley owned.

Then I looked into the proposition seriously. I went over the ground in person, realized the vast opportunities presented, particularly in the great forests of the Eel River country, which were still government land. The way things were going then, it would have been no trick at all to introduce a bill in Congress asking for a land grant through a country to be traversed by a railroad and get half a million acres or more just for the asking. It seemed to me a bigger game than all the gold mines, speculations, and investments I had ever seen or dreamed of. I tried to interest Ralston, but he said I was visionary and made some remarks about "back lands" and "coyote ranges."

That did not deter me in the slightest. I had abundant capital of my own and very important financial connections and had no doubt that I could complete the undertaking on my own account. With a good corps of engineers I began to rush the work of surveys and locations with my customary impetuosity. In a short time I had the dirt flying at Petaluma and several other points north. I contracted for fifty miles of ties as a start and bought fifty miles of rail, some ten miles here and the rest in England. I was perfectly infatuated with the railroad business and determined to devote my life and energies to the work.

Needing all the money I could get to handle this enormous enterprise, I suggested to Ralston that we hold an auction sale of our joint possessions. We had laid out New Montgomery Street in good style. We had completed our plans for the Grand Hotel, and the inquiry for our holding was brisk. Besides, San Francisco was in the grip of a tremendous real-estate craze, the biggest in its history. The railroad would be with us in a few months. Then everybody would be rich.

We had incorporated under the name of the Montgomery Street Land Company. The moment an auction sale of its properties was announced the whole town was alert. The offices of the company were crowded with investors eager to purchase at private sale, but were told that we were going to have an old-fashioned auction and nothing else.

It was less than a week before this historic event took place when the minimum prices were arranged. Ralston, Maurice Dore, and myself met in a back office of the Bank of California one night and discussed this all-important question. Finally we agreed that each should write on a slip of paper his opinion of an average price per front foot. I based my figures on a profit of two and a half millions, which seemed to me a fair return. But when we came to compare the slips, Ralston's figures were just double mine, while Dore's were intermediate—nearer mine.

Ralston's nature was sanguine. He never saw anything but success. He had supreme confidence in his judgment, not without foundation, and possessed a knack of bringing everyone to his own views. If he was right in this instance, of course, five million dollars were more desirable than two and a half. I yielded to his arguments, but not without grave misgivings.

That auction sale was memorable for many a year. By consent, it was held on the floor of the Merchants' Exchange, and there never was such a throng of moneyed men gathered together in San Francisco. Everyone seemed keyed up to buy a lot or have a free fight. Maurice Dore and his spieler, Cobb, were past masters in all the auctioneer's arts to promote enthusiasm. Among his "cappers," to bid the prices up, were Mills, Hayward, Sharon, Tom Selby, and William Alvord. Pandemonium broke loose when the first offering was announced. Men fought and raved, like brokers filling "shorts" on the stock exchange. The same scenes were re-enacted time after time, but it became only too evident to insiders that our "cappers" were picking up everything ostensibly sold. The fact was that the public would have gone above my estimate, might have touched Dore's, but stopped short of Ralston's. After keeping up the hippodrome long enough to save our faces, the great sale was adjourned without a day.

But that wasn't the worst of it. For months we had been living in a fool's paradise over the boom that would follow the driving of the last spike. That day came, but what a disappointment! It may have seemed all right to the proletariat, but for the business people it spelled ruin. It brought in an avalanche of goods from Chicago and St. Louis at prices which our local men could not meet. Many firms failed, some consolidated, some retired from business. Rents dropped like lead, real-estate values shriveled up to nothing. It was ten years before those values recovered to the level of 1869.

Meanwhile, my railroad in Sonoma was being rushed ahead. I do not think it would have encountered opposition had I stuck to my original plans of a coast line through Marin, Sonoma, Mendocino, and Humboldt. But my vision began to broaden. I knew of Beckwith Pass and the almost increditable fatuity that overlooked it in constructing the Central Pacific Railroad. I employed General W. S. Rosecrans and a corps of engineers and began a railroad survey westward from the pass, to connect with the Humboldt system. I guess this gave the Big Four the largest scare they received in many years.

Instantly I found my enterprise blockaded with all kinds of petty obstructions. I had thirty miles of road graded and the ties strung. Peter Donahue had partly agreed to sell me the rails. Suddenly he withheld them from the market, and there were no more on the Pacific Coast. I had twenty miles of rail on the water from England. The vessel was detained at Valparaiso and at last sunk peacefully in the harbor. My agent, A. A. Cohen, always claimed the ship was scuttled.

I ordered another large shipment from England. Then the railroad people tried another tack. They appealed to Ralston to subdue me. Ralston had been the friend of the Big Four in the trying construction days. They had promised him all sorts of things in return, among others a concession to build all their cars, for which he made great preparations. They plainly told him that if he did not constrain me, his estimable partner, to abandon my railroad projects, the concession would be canceled and he could expect nothing but war.

Ralston laid the matter before me as a friend. He admitted that he had no right to influence my action but said he was facing an enormous loss, that I could sell out at a large profit, and frankly asked me to strain a point. The matter once placed in that light, I yielded, with great reluctance. After some negotiations, I sold my railroad rights to Peter Donahue. These rights, only partially developed, constituted the bulk of his great fortune.

That incident made a vast difference, not alone in my fortunes, but in the history of California.

Left to myself, I would have had a railroad to Humboldt Bay thirty-five years ago and would now be the owner of the Northwestern Pacific Railroad.

Also, it is more than probable that my youthful energies would have carried another railroad eastward through

Beckwith Pass. That would have made history, changed our governors, United States Senators, bosses, and the whole machinery of state.

As for poor Ralston, he was gold-bricked. He never received the car concessions at all. I cannot tell why, for I was out of the state when the scheme went up in smoke. The great building he had constructed for the purpose was converted into the West Coast Furniture Company's plant, which was operated during his lifetime at a heavy loss.

I can only think of Ralston as a long-cherished and lamented friend. But so far as business went, our acquaintance began and ended under an unlucky star.

CHAPTER XIX

Two Men Block Plan to Run New Montgomery Street to the Bay; One Asks Coin, Other Prefers Fight

Promoters Appeal to Legislature and Do Not Neglect Precaution of First "Seeing" Vote Brokers

When Ralston and I opened New Montgomery Street, we never doubted that its manifest importance would compel an immediate and voluntary extension to the natural terminus of the water front and prove the logical outlet for congested trade. That this would have been the case had the majority of property owners been able to follow our example, I have no reasonable doubt. But just as in the case of "Montgomery Street Straight," special interests and selfish considerations stood in the way. Less than half a dozen property owners, to their irreparable disadvantage, blocked "Montgomery Straight"—a project that would have changed the whole course of the city's progress and development. Just two property owners prevented the immediate extension of New Montgomery Street to the bay, and again the failure was the city's heavy loss.

These two men were Milton S. Latham and John Parrott. Latham owned a stately home and large grounds on Folsom Street, directly in the line of the new thoroughfare. It was a matter of no small personal pride, and doubtless he was attached to the locality. He asked such a fabulous price for the right of way, which, of course, would have destroyed the home value of the property, that even Ralston and myself, who were accustomed to brush any minor obstacles out of our way without counting costs, stood aghast.

John Parrott, on the other hand, wouldn't trade at all. His business hours were then strictly limited from nine to half-past ten, and every time we managed to secure an interview, all the satisfaction we could get out of him was a promise to fight us every inch of the way.

Outside of these two, we had a clear field. We secured contracts on a great number of properties along the line of the proposed thoroughfare. All the large owners concerned favored it with enthusiasm. Still we were absolutely blocked.

Under these conditions nothing remained but an appeal to the Legislature to appoint a commission, empowered to open New Montgomery Street for its full length and assess benefits and damages as provided by the general laws then in force.

And while about it, we did not stop there. We worked out a grand, comprehensive scheme of improvement, embracing the immense territory to the south.

Two years before, a bill had been lobbied through the Legislature providing for what became famous later on as the "Second Street Cut." It was a rascally project, a sordid bit of real-estate roguery, carried through without a moment's thought of other people's rights. But it was an accomplished feat, and one of the results was to ruin the finest haunt of good breeding San Francisco ever had. Families were scattering from Rincon Hill to various sections of the city. The old high-priced residence property was going for a song. As the "Hill" had ceased to be either beautiful or useful, Ralston and I calmly proposed to cut it down.

We planned to have the city buy the property, which could be purchased for $5,000,000 according to arrangement with the owners, and grade it to the Market Street level. Many million cubic yards of excavated material were used to fill in a 150-acre tract of tideland, offered to the city by the state at a nominal price, lying between the Pacific Mail docks and Islais Creek; also to reclaim China Basin, at least in part. The cost of grading and reclamation work was estimated at $7,000,000; in fact, contractors were willing to undertake it at that price. In other words, the city was asked to issue its bonds for $12,-000,000 and receive in payment over 200 blocks of choice property, to say nothing of great advantage to the appearance of the town and the facilities for doing business.

Two separate bills were introduced in the Legislature. One provided for the opening of New Montgomery Street to the bay and created a commission to carry out its purpose as above defined. This would probably have slipped through without any serious opposition, but coupled with it, in a way, was the great construction bill for acquiring Rincon Hill, for filling the tideland acreage and China Basin, and running all the streets from First to Third, including an extension of Sansome Street, on a nearly level grade, southward to the water front. For the extension of Sansome Street, Michael Reese, Lloyd Tevis, and myself had bought a solid block from Market to Folsom Street.

I was very much a novice in politics, but Mr. Ralston insisted that I should have full charge of the program and take up my residence in Sacramento, pending the session of the Legislature. So among other things I gathered quite an exact idea of how wires used to be manipulated underground.

In the first place, the necessity of a legislature was not apparent at that time. What had been an able and independent body in the early history of California had degenerated to a mere recording machine for a couple of vote brokers, "Nap" Broughton and "Zeke" Wilson by name.

"Nap," brief for Napoleon, was a happy, enthusiastic chap, always slapping someone on the back with a heartiness not always quite sincere, a good fellow in his way, and a most abandoned corrupter of men, a spendthrift disciple of nearly every sin, with an ever-watchful eye on the money of others, yet himself the veriest sucker that ever lived.

"Zeke" Wilson, on the other hand, was a gray, desiccated, sinister old spider who seldom smiled, and when he did, everyone in his presence felt depressed. He was the "thinking member" of the duumvirate, and while "Nap" Broughton made nearly all the noise, "Zeke" Wilson laid the plans.

The Senate used to be respectable in appearance, an able body, and reasonably clean. The one that I was concerned with contained such men as Hager and Saunders of San Francisco, George C. Perkins of Butte, who made then his first appearance in politics, Rumbaldo Pacheco, afterward governor, Pendergast of Napa, Lewis of Tehama, and several others whose names are fairly connected with the history of the state.

The Assembly, on the other hand, was a conglomeration of miscellaneous riffraff, gathered together God knows how, inexperienced, ignorant, venal, and scandalously cheap. Of course there were some honorable exceptions. I am only speaking of the general rule. It was in the Assembly, not the Senate, that the "business" of the session was done. That is, if Messrs. Broughton and Wilson wanted to kill a measure, they never worried what the Senate did, but let the obnoxious bill come before the "popular-priced" Assembly, where its shrift was short.

No one in his senses ever came to Sacramento with a bill involving a considerable question of finance without establishing friendly relations with Messrs. Broughton and Wilson at the start. Treaties of alliance were negotiated through Napoleon Broughton. At our first interview, $35,-000 passed hands. "Nap" merely said in a casual way that I was a gentleman, and I accepted the compliment for what it was worth. What became of that money I have no means of knowing and never inquired. That would have been the height of bad manners. But he never asked me for any more, and everything I wanted slid through the Assembly on greased ways.

We were among the first who made a consistent effort to impress the merits of our measures on lawmakers by systematic good fellowship. I practically chartered a well-known restaurant, threw it open to my friends, and the bills were over $400 a day, so generously did they respond to my invitation. Down in San Francisco, Ralston was on the lookout for statesmen, and none of them struck the town without good cause to remember the experience pleasantly.

In a way, it was a striking session—a sort of breaking of new ground. The railroad appeared for the first time as a seeker for favors. It had two leading bills, each providing for a subsidy for railroads southward, one through the San Joaquin Valley and one along the coast line. Neither terminated anywhere in particular, the former somewhere in Kern County, the latter in San Luis Obispo County near the border line of Santa Barbara. The measures simply authorized the electors of the counties concerned to vote for a subsidy payable to the first railroad that came along. The combined subsidies provided for amounted to only $3,000,000, but they were regarded as the opening wedges for more. Of course everyone knew what that first railroad would be. Strangely enough, in the newspaper and legislative discussions, no one seemed to think that Los Angeles cut any figure as a terminal or feeder. The cry was for a railroad south to the Colorado River. For that the people were willing to pay any kind of subsidy, but not a cent for a couple of local concerns. A bitter newspaper war followed, and charges of corruption were freely made. But the bills passed both houses by large majorities and were only halted in their triumphant progress by the veto of Governor Haight. Even then it was a close call. The Assembly enthusiastically passed one of them over his veto, and in the Senate the same action failed by only two votes.

There were so many bills of a shady, not to say rotten, nature introduced during the session that almost all measures were looked on as "jobs." Our two bills—"Montgomery South" and the effacement of Rincon Hill—took their places with the rest. They were harshly criticized by most of the San Francisco papers as crafty schemes, the true inwardness of which would develop later on. They were likened to the "Second Street Cut" outrage, and a lot of ill-advised public opinion was worked up against both. Nevertheless, they passed the Legislature. How one of them became a law is an interesting story, told in many official records of the state.

The bill for the extension of New Montgomery Street had gone to Governor Haight. It leaked out from the executive chambers that a veto message was being prepared. The Governor had ten days in which to veto the bill; otherwise it became a law by default. It was on the afternoon of the last day, shortly after the Senate had reassembled, when one of my attorneys, Creed Haymond, said in a musing way, "If the Senate could only be induced to adjourn, we would not have to worry about a veto message. Then it could not be delivered to anyone and by twelve o'clock tonight would become a law." That set me thinking in a moment. "Is that correct" I asked. Haymond replied that he was certain, although he was not sure that the point had ever been tested by the courts.

The emergency demanded swift work. To offer a motion to adjourn just after settling down to business would certainly have aroused suspicion and a general rumpus. Here I worked in a bit of strategy, or what might have more properly been called chicane, which I trust may be pardoned me in my final account.

Senator John S. Hager was the leader of what might be called the "reformers" and had quite a following among his fellow members. He was the unwearied foe of anything like a job. Among other measures, he had opposed the Montgomery Street Extension bill. But there were several bills on a special file that afternoon that were his pet abominations, and he justly feared that they might slip through. While in this frame of mind a certain gentleman called him aside and advised him that several members were anxious for an adjournment, that if he would make the motion, it would probably carry and the obnoxious bills would lose their places on the special file and their chance of final passage.

The Senator swallowed the bait—hook, sinker, and all.

While he was lining up the "reformers," somebody else was attending to the "performers," and when the gentleman made his motion to adjourn, he must have been gratified at the unexpected support. It went through *nem con,* as the lawyers say. The officers of the Senate were hurried out of the room on one pretext or another, and in a few minutes the chamber was vacant.

Dr. Edward R. Taylor, later mayor of San Francisco, was the very efficient private secretary of Governor Haight. I was in an agony of fear lest he should pop into the Chamber with the fatal message before the adjournment could be arranged. For this reason I had several effective conversationalists stationed between the Governor's office and the Senate to engage the secretary for a few minutes if he chanced to appear. This they actually did, although Dr. Taylor has forgotten the incident. What he does remember was that he found, much to his surprise, an empty Senate Chamber, and after ruminating over the situation for a time, carried back the veto message to the Gorvernor's office and laid it on his desk.

On the following day the Governor attempted to deliver his message, but the Senate held he was too late. His Excellency refused to certify the bill to the Secretary of State as passed, and I brought a mandamus suit to compel him to take that action. The title of the case was *Harpending* vs. *Haight* and attracted a wide attention at the time. It was carried to the Supreme Court on an agreed statement and decided within fifteen days in my favor. The decision can be found in volume 39, *California Reports,* page 189. Other governors have been cautious not to hold back their vetoes till the last day. Hager roared like a wounded bull buffalo when he found out how he had been used, but his lamentation bore no fruit.

Thus the Montgomery South bill became a law of the state, although the Governor liked it not. Commissioners were appointed by Judge Lake, a lot of work was done in surveys, estimates of benefits and damages, but in the end it came to naught. Two years later, while I was in Europe, a bill with a misleading title, designed to repeal the act, was introduced, and Ralston, busy with many things, never knew about it until it had sneaked through both houses and become a law. Because of this, New Montgomery Street still halts at Howard Street and bids fair to camp there forevermore.

As to the Rincon Hill measure, that also passed both

houses triumphantly, but was held back through the opposition of Senator Hager, so that it went to the Executive just one day beyond the period when a return to the Legislature must be made. It found a peaceful resting place in the Governor's capacious pocket.

Thus all our grand schemes for the development of the city southward fell by the dreary wayside of lost opportunity. I do not pretend for a moment that Ralston and myself were inspired in our efforts by the pure spirit of benevolence. We would have made our profit, but a mere trifle in comparison to the public good. It was the most comprehensive plan for the city's improvement ever presented in a concrete form, and the pity is it was not better understood.

Just take a retrospect. Who is there who would not admit that five fine level streets from Market, between First and Third, southward to the bay, would not be a vast improvement and convenience to business, over the blockade that prevails today?

And was such a real-estate proposition ever before offered to a people and turned down? For the sum of $12,000,000 the city would have acquired full title to approximately 220 blocks, the present value of which would be hard to estimate exactly. But a rough valuation indicates that the property would be worth enough to pay the entire city debt, buy the Spring Valley Water Company's plant, bring in the Hetch-Hetchy water supply, and leave a balance large enough perhaps to settle all questions with the United Railroads and municipalize the entire street transportation system, not in the dim future, but now.

Immense revenues would have flowed into the municipal treasury from these utilities. Taxation would have become a joke. All these things are among the haggard, melancholy "might have beens."

There were too many well-intentioned, but bigoted, reformers in the city then, just as there are now.

And the incident serves to indicate the superiority of hindsight over foresight, which has been illustrated unhappily and too often in the history of the state.

CHAPTER XX

Burning of Harpending Block Provides Fine Spectacle, But Oversight of Owner Costs Him Dearly

George Hearst Makes Stake on Comstock and Celebrates by Taking Joe Clark on a Trip to Europe

I WAS BUSY with other things besides real-estate investments, financing railroads, and politics during the five years between 1865 and 1870. In 1869, I built the first fine business block on the south side of Market Street, the Harpending Block, between First and Second streets. It was also in 1869–1870 that Ralston and myself built the Grand Hotel, partly on our own land, partly on land belonging to the Platt Estate, which we held under twenty years' lease.

The Harpending Block cost nearly $400,000. It was burned two years later, contributing the biggest fire in San Francisco since the fifties. Through an oversight of my agent, the insurance hardly represented a tenth of the loss. The Grand Hotel remained for several years the last word in the hotel business of the Pacific Coast. It was its phenomenal success from the onset that induced Ralston later to embark in the Palace Hotel project, which contributed in a large measure to his ruin. I owned a three-fourths interest in the Grand Hotel; Ralston owned the balance.

No one who has ever had much to do with mining can keep out of that fascinating business very long. When I returned to San Francisco from Havilah, it was my solemn intention to abandon mining forever thereafter and confine my efforts to what was known as "legitimate business";

whatever that may be, I have never found out. But I hadn't more than barely got my bearings before I began to make casual incursions in a sly way into the old field of endeavor and thus had a personal and financial acquaintance with, I think, all of the heroic figures who created the vast deep-mining industry of the Far West.

Only one of these big men has lasted down to our own time. J. B. Haggin[1] still lives at his home in Lexington, Ky., at a great age—ninety or more—and until recently in the enjoyment of good health. A story used to be current in San Francisco that in the early pioneer days, Haggin was a devotee of play at the El Dorado and Union. One night, so the narrative runs, after successive losses, he borrowed $100, to win or take the gambler's last alternative. But he had no occasion for the latter. He stood calm and imperturbable as the hundred became a thousand, and then tens of thousands, while a circle of mute white-faced gamblers stood fascinated at his luck until the proprietor, in a voice that showed no tremor, quietly announced the bank closed for the night. The story goes on to say that Mr. Haggin never touched a card from that day forth. All of this I have only on hearsay. Mr. Haggin lives to tell whether it is true or false.

But if he abandoned gambling in one direction, he took it up in another. In the mining industry he was a plunger, par excellence. I do not mean that he invested recklessly or without mature investigation, but when he once made up his mind, a few millions, more or less, never moved him from his purpose. The broad, liberal way he played the game had more to do with the development of the West than perhaps anything else.

Haggin had nothing in common with good fellowship. He was always silent, sober, and cold. But under it all he

[1] Since the above was written, Mr. Haggin has died.

J. B. HAGGIN

Successful miner and a true financial genius

must have had a heart. He was the only one I ever knew who remembered the men who helped to give him wealth. Every man, without exception, who rendered Haggin faithful, efficient service he made rich. And he was very loyal to his friends. In these days—and other days—when men of power exhaust the energies of their subordinates and then toss them without concern on the scrap pile like so many sucked-out oranges and treat their business associates just a shade better, an example such as Haggin gave ought not to be overlooked.

George Hearst was probably the greatest natural miner who ever had a chance to bring his talents into play on a large scale. He was not a geologist, had no special education to start with, was not overburdened with book learning, but he had a congenital instinct for mining, just as some other people have for mathematics, music, or chess. He was not a man of showy parts, liked the company of a lot of cronies, to whom he was kind and serviceable—when he wasn't broke himself—was much inclined to take the world easy, but if anyone mentioned mines in his presence, it had the same effect as saying "Rats!" to a terrier. Hearst became alert and on dress parade in a moment.

Hearst made his first big stake on the Comstock Lode, a year after it was uncovered in 1858. He was associated with his cousin, Joe Clark, and William M. Lent. I do not know the exact size of the cleanup, but it must have reached into seven figures. Such an event in the old days was always made memorable by some kind of a "jamboree."

Now Joe Clark was a southwestern man, hailing from a section not far from where I originated myself. All of us were inclined to be provincial. For instance, Joe Clark believed that St. Louis was not only the most magnificent, but the largest city in the world. He had many heated discussions on the subject and several times backed his opinions with coin. He declared that the Rue de Rivoli was a pale shadow alongside of the glories of Laclede Avenue. He swore that St. Louis was bigger than London, more cultured than Athens during the age of Pericles, and grander and more picturesque than Babylon when the hanging gardens were in full bloom.

It is said that Hearst suggested a "blowout" in Europe after their cleanup, in order to disabuse his kinsman's mind of certain illusions respecting St. Louis. At any rate, the two husky young miners set their faces eastward to look over the effete monarchies of the Old World.

While they were pleasure bound, "Bill" Lent stayed behind to look after the investments. He sunk a shaft which headed dead on for the big bonanza, and had he continued the work a little further, Flood, O'Brien, Mackay, and Fair would have cut a very small figure in history. But he engaged, unfortunately, in a seductive-looking speculation and went to pieces in a grand, pyrotechnic, and spectacular failure. Hearst and Clark were hopelessly involved. They received the news while they were making the tour of Europe with much éclat. Fortunately, they had money enough to reach home. But the main object of the journey was accomplished. When Joe Clark mentioned St. Louis thereafter, it was the voice of a chastened soul that spoke.

Of course, nothing could keep Hearst down in a mining region. Any capitalist was only too eager to back a man with such surpassing talents; but he had to pay an awful toll. For years, Hearst's projects were financed at 2½ per cent per month, compounded monthly, and any business that can stand that strain and come out ahead must have a solid foundation to build on. He was the real founder not only of his own, but of the vast Haggin and Tevis fortunes.

I had mining deals of more or less importance with Haggin, Hearst, Hayward, Hobart, Grayson, in fact, with nearly all the large operators of those times. My largest speculations, however, were with Ralston as a silent partner, which, on average, showed more profit than loss. It was for the purpose of joint investment that late in the fall of 1870, I visited the Emma mine near Salt Lake City, which a year later was the central point of a great international scandal and will play an important part in this narrative.

CHAPTER XXI

SAM BRANNAN STRIKES IT RICH AND REFUSES TO SHARE WITH MORMON CHURCH EXCEPT ON ORDER FROM LORD

Mine Bargain Fails to Stand Inquiry of Author, But Others Invest and Figure as Victims of Fraud

I HAD EARLY BEEN familiar with Utah and its mines through an acquaintance with "Sam" Brannan. Brannan had a history of thrills and adventures which, if gathered into book form, would make the heroes of Dumas look cheap and commonplace. Originally a Mormon, high in the councils of Brigham Young, he led a body of his co-religionists around Cape Horn to California before the earliest Argonauts. He staked out claims on the American River, about two miles from where Folsom Prison stands, the location being known as "Mormon Island" to this day. The diggings were so rich that one of California's evanescent cities sprang up around it almost overnight, just as suddenly to disappear. "Sam" worked his companions on a per-diem basis and very soon accumulated a large fortune—certainly in excess of a million dollars, many well-informed people estimating it at two or three times as much. But while he settled promptly his labor bills, he was not so businesslike in squaring accounts with the Mormon Church, which claimed nearly all the profits. Finally, a trusted agent was dispatched from Salt Lake City with a peremptory order on Brannan to turn over the ecclesiastical share of the "dust" at once.

Brannan's reply was historic and to the point, even if a bit profane. The gold, he said, had been placed in his safe keeping on the Lord's account. He would surrender it upon the Lord's proper written order; otherwise not.

"Sam" invested most of his wealth in San Francisco real estate. An important street bears his name. Like most of the early Mormon leaders, he was of a coarse-fibered nature, with a rather forbidding, saturnine face, but singularly keen witted, resolute, and fearing neither man nor devil.

The latter quality stood him in good stead. Brigham could not permit such a flagrant breach of church discipline to remain unpunished. Flock after flock of "destroying angels" took flight from Salt Lake City, duly commissioned to bring back Samuel's scalp or perish in the attempt. But their holy work was always a dismal failure. Brannan must have had some foreknowledge of their movement against the security of his person. Liking not to meet "angels" unawares of any kind, he arranged to encounter the "destroyers" halfway out in the trackless desert, or mountain fastnesses, with a competent group of exterminators he seemed to keep on hand for such occasions, and it was the "angels" who were always taken unawares. Some of them got back to Salt Lake minus tail feathers and otherwise damaged, but the majority of them never returned at all. At last the disciplining of Brannan became so manifestly an extra-hazardous risk that it was finally abandoned. How he defied the whole power of Mormonism and actually conducted a private and successful war against the church was one of the old romances of the Pacific Coast. In later years Brannan fell a victim to drink, all his enormous wealth became dissipated, and he died, penniless and forgotten, in Mexico.

"Sam" never forgot Salt Lake City or Utah. His life would not have been worth ten cents if he had once stepped within the territory of Brigham Young. But he always cast longing eyes at the scene of his early struggle. He knew

Utah and its resources from end to end and in our frequent interviews often mentioned the illusive "pockety" nature of its mines. Therefore, when Ralston and I took a thirty-day option on the Emma mine, about forty miles from Salt Lake City, I was prepared to exercise extreme caution in examining the property.

The Emma mine had startled the Coast with a wonderful burst of production, considering the limited nature of its plant. Its wealth was claimed to be fabulous, and it was a matter of surprise when its owner's offer to sell it at the low price of $350,000 was made. Nevertheless, the proposition seemed worth looking into. But, remembering Sam Brannan's counsel, I went unannounced to the mine, presented my credentials to the superintendent, who gave me permission to examine the property, although rather surprised that I came alone.

It did not take me long to reach a conclusion that the Emma mine was nothing more than a large "kidney." Considerable high-grade ore had been stoped out of the upper levels. Below, the ore was plainly pinching out. The whole thing was nothing but a shell, with just enough in place to fool a tenderfoot. There was no trace of a fissure vein. Any mining expert would have turned it down without a moment's hesitation.

I had just seen all I cared to see when J. W. Woodman of Salt Lake City, the principal owner of the property, hurried to the mine in some agitation and expressed his regret that I had not advised him of my coming. However, he trusted everything was satisfactory. I told him courteously that I could not pass favorably on the mine and to consider the option closed. He wished to argue the matter, but I told him that the conclusion was final and decisive. Then he took another tack. He was anxious, he said, to clean up and get away. If he threw off an even hundred thousand dollars, would Mr. Ralston and myself take over the property? Again I answered in the negative and told him, in so many words, that we did not want the mine at any price.

As a matter of fact, when it was finally bottomed, the mine did not yield anything like a hundred thousand dollars net. I even doubt if it made both ends meet.

Such was the Emma mine, famous, or rather infamous, in history. Just a little later this barren hole in the ground figured in one of the biggest swindles of modern times, in which great names were involved, a minister of the United States to England disgraced and ruined, British investors robbed out of ten million dollars, and the business world filled with such suspicion that for many years the doors of foreign credit were barred against American mining enterprise of every sort. The very character of Americans for common honesty was so seriously besmirched that it caused an international unfriendliness that time only partially cured.

It was a fact of common knowledge that the mine had been bonded to Mr. Ralston and myself for $350,000. It was also well known that I had examined and must have found it unsatisfactory, for the bond was allowed to lapse. This alone gave the Emma such a black eye on the Pacific Coast, then the great market for legitimate properties, that it became almost a waste of time to make any further attempt to market it in the neighborhood where the circumstances were known. Ralston was rather noted for taking a long chance on mining ventures, and while his luck lasted, he usually pulled them through. Therefore, when he and his associates turned down a developed and going concern, the wise, conservative natures shook their heads. That is doubtless why the Emma was taken to a market some five thousand miles from home for exploitation.

In fact, it was practically taken off the market for quite a while. After it was first offered for sale to Ralston and myself, my impression is that I was the only one who examined it qualified to pass an honest judgment on such a property, until it suddenly blossomed on the London stock market as the great American ophir, the newly discovered treasure store, of which the human imagination had dreamed for ages—and was unloaded on the British public for $10,000,000 or, to use the parlance of our Anglo-Saxon cousins, for £2,000,000.

I have gone into the early history of the Emma mine so minutely because it strikes this narrative a little later at an angle so acute that the two seem to run parallel, and it is necessary to have all the facts in hand to understand how the great swindle that strained the commercial friendship of two great peoples almost to the breaking point had a close relation to the diamond hoax story.

CHAPTER XXII

Briton with Oriental Imagination Seeks to Lure Investors with Tales of Mountain of Silver

New Promotion Company Tells Truth, But Editor Samson Frightens Off Public at Critical Moment

When I reached Salt Lake City after examining the Emma mine, I found awaiting me a telegram from Mr. Ralston to the effect that the president of the Bank of England, a Mr. Green, then traveling in the Far West, would be in Cheyenne on a certain day. He asked me to meet the gentleman and in his name, as president of the Bank of California, extend to the visiting banker any courtesies that his time and inclination might permit. So I journeyed to Cheyenne in quest of Mr. Green.

I stopped at the principal hotel, and one of the first persons my eyes rested on was about the most impressive-looking man I ever saw. He must have been six feet six in his stocking feet; he was richly caparisoned, handsome, debonair, evidently a Briton, and looked like the president of the Bank of England and the Prince of Wales rolled into one. I took a chance, approached the stranger, and asked him if he were Mr. Green, president of the Bank of England. The gentleman laughed and said I had made a close guess but had missed the mark a trifle. He introduced himself as Mr. Morgan, an Englishman of leisure, making a sight-seeing tour of the Far West. Later I discovered that Mr. Green had passed on without stopping and was then well along on his journey east.

One of my objects, besides inspecting the Emma mine, was to examine a property I had acquired in New Mexico near the headwaters of the Gila River. I had made an investment on the strength of huge outcroppings of mineralized ledges that gave indications of a great mining property. But besides that, there was a large valley, covered waist deep with grass interspersed with black walnuts into which luxuriant wild hops twined and traversed by a fine stream of water. In addition to the mining claims, I had secured the water rights and taken the preliminary steps to acquire a vast acreage of fertile land. Development work had been going on for some time, and I was anxious to see for myself just how the property was showing up. I had several chats with Mr. Morgan after our first odd meeting, and, learning of my projected trip to New Mexico, he asked and readily received my consent to go along.

Arrived at our destination, Mr. Morgan at once became infatuated with the country—ledges, land, water, and all. Some of the prospect work showed ore of high values. The Englishman took many samples and had them tested by my assayer. My impression is that like every beginner in the mining business, he always chose the best. Finally, he made me a business proposition. He said he had important financial connections in England, that a great diversified property like this could be floated for an immense sum—named $3,000,000 as a fair estimate, and offered to form a company on an equitable basis to finance and develop its resources.

With a cooler head I advised Mr. Morgan that the mines were still only in the "prospect" state, that they might turn out something great, but more likely nothing at all. Concerning the land and water there was no question. Properly handled and developed, their value must be great.

After some negotiations, we hit upon a bargain. Morgan was to go to England post haste. I was to follow by more leisurely stages a month later, and by the time of my arrival everything was to be arranged.

I stopped a few days in New York to see the sights. While there I met another Englishman by the name of Dalton, a member of Parliament. I told the gentleman something of my contemplated trip to England. When I mentioned the name of Morgan, he seemed a bit amused. He said Morgan was all right, that he had excellent family connections, but that he hardly figured as a financier. He said that his imagination was of an Oriental type, prone to exaggeration, and very apt to make a mess of any large transaction. "If Mr. Morgan fails," he said, "you had better come to me."

When I arrived in England, I found that Mr. Dalton's prediction had already come true. Morgan had issued a prospectus that put the tales of Baron Munchausen in the shade. He actually described the mines as mountains of silver and by his very extravagance of statement doomed the enterprise from the start. Meanwhile, I had various meetings with Mr. Dalton, who was a man of standing in the business world, and through him met a great firm of brokers, Coates and Hanky. Mr. Coates was the son of a manufacturer who won fortune and immortality by his exploits in spool cotton. These gentlemen agreed to place my proposition before the investing public. Morgan floundered around for a short time but was soon discouraged. I offered him an interest in the new exploitation, with the understanding that he keep mute.

Coates and Hanky now understood the enterprise in a business fashion. The New Mexico Land and Silver Mining Company was formed, with a high class directorate. One of the directors, I recollect, was a retired admiral of the British Navy. The prospectus was flattering enough but would stand investigation. Among other things, it dwelt more on the unquestioned value of the land and water than the probabilities and possibilities of the mines. The capitalization was £600,000.

The London *Times* was then, as now, the great newspaper authority of England. Its financial editor, whose suggestive name was Samson, was currently said to have more power than the Queen. Five lines favorable from Samson's pen in the financial columns of the *Times* assured the success of an enterprise. Five lines unfavorable were equivalent to a death warrant. It was customary with promoters to submit their plans to Mr. Samson before submitting them to the public. The directors of the New Mexico Land and Silver Mining Company followed this custom and received a somewhat cryptic answer which, however, they construed to be favorable.

The issue was brought out with the skill of trained hands. Everything pointed to a successful outcome. But the very next day, Samson came out with a double-barreled blast. Before the *Times* reached the country, a small avalanche of subscriptions poured in. But in the city, after a large first day's business, the promotion fell flat. Nevertheless, the directors stood manfully by their guns. They received space in the *Times* to answer. They put up a bulldog sort of fight. The old admiral in particular was as belligerent as when he paced a man-of-war. There was somewhat of a reversal of public opinion in our favor. More than half the capital stock was subscribed for, and we might have pulled the issue through, but it seemed to me that the company was overburdened to start with, that it must labor under too many handicaps of distrust to operate successfully, and against the judgment of the directors I withdrew the properties and the incident was closed. All the subscribers received their money back, without cost or abatement. No investor lost a cent.

An incident shortly after my arrival served to illustrate in a pleasant way my relations with W. C. Ralston at that time. I was asked to call at the Oriental Bank, the agency for the Bank of California, and, going there the following day, was ushered into the presence of the president, an impressive-looking man of affairs. "I have here," he said, "a cable from W. C. Ralston, president of the Bank of California, advising us to give Mr. A. Harpending credit for any sum he wants. This is an unlimited order, and as you probably intend to make heavy drafts on us, I thought it advisable to inquire beforehand how much you were likely to want." I laughed and told him I had all the money I needed, but if I happened to want accommodation, I would certainly call for more. The story is immaterial in itself, except as an illustration of Ralston's offhand way of doing business and his confidence in me as his friend.

Another pleasant incident was the renewing of my acquaintance with Alfred Rubery, who again becomes a leading figure in this story. He was the same old Rubery of the "*Chapman* days." John Bright, his illustrious uncle, was at the height of his prestige and power, and Rubery himself was in the swim with the biggest kind of social and political fish.

And still another incident was that I came in personal contact with the famous Baron Grant, the overlord of financial London.

CHAPTER XXIII

Baron Grant Demonstrates His Talent for Exploitation by Putting Over a Deal that Nets $1,500,000

Happy Directors Decide that Occasion Calls for Generous Cash Souvenirs, But Stockholders Object

Those who are familiar with the staid, conservative, even-paced London of today can hardly realize what the same London was in 1871, the period of my first visit there. It was the year of the great Franco-Prussian War. The pleasure capital of the world was transferred from the River Seine to the River Thames. Male and female adventurers of every nation thronged the British capital; speculators eager to tap the great reservoirs of English wealth, gentlemen who lived by their wits, *chevaliers d'industrie* in general made London a common trysting place. And the life was to correspond. It was notable for undisguised and shameless intemperance, a primitive, savage, heathenish pursuit of women, and a fevered spirit of gambling speculation that cut loose from all moorings of common sense. I could compare it only to the recklessness and abandon of a Western mining camp in the orgy of flush times.

The speculative world was ruled and controlled by a strange character, for many years one of the famous figures in London, Baron Grant, the same man I mentioned in the last chapter. He was half-Hebrew, half-Irish, and it has been my experience that wherever you find that combination, you can look for something different from the common run. His real name was Gottheimer, but he had it changed by act of Parliament to Alfred Grant. He came by his title in a curious way. When the nascent kingdom of

Italy, years before, had attempted to raise a considerable sum and had been turned down in the money marts of Europe, Grant, then in the height of his prestige, offered his services and floated triumphantly the discarded securities, for which service the grateful Italian government honored him with the title of baron.

When I first met him, Baron Grant was past his zenith. Some of his transactions had been disapproved by the great financiers, but he was still a potent factor in the domain of speculation and a promoter without peer.

Personally, he had the magnetic temperament more highly developed than any man I ever knew. His manners were engaging, he was simply a wonder in conversation, and as he spoke, his handsome face was lighted with candid smiles that no one could resist. Whoever came within the sphere of Baron Grant's influence felt the intoxication of his power to charm.

Meeting several times under favorable auspices, we talked of the mines of California and the trans-Mississippi region in general, concerning which I could speak with first-hand knowledge. He was deeply interested, said that such properties would have a ready sale on the booming London market, and promised that if I could only secure an option on a high-grade mining proposition, it would prove a very profitable piece of business to both of us.

I cabled Mr. Ralston, naming three well-known developed mines and asked him to secure me an option on one of them. In answer I received a cable from William M. Lent, president of the Mineral Hill Silver Mining Company, in which I owned a quarter-interest myself, offering an option on that property for a million dollars. Within a month all the necessary papers arrived by mail. These included, besides a legally drawn option, a full description of the property, its productive history, maps, engineer's reports, estimates of tonnage in sight, and all the details that a careful investor might require. In addition, there was a private agreement, duly executed, giving me a commission of 10 per cent.

It certainly was an alluring proposition. The Mineral Hill mine was located in eastern Nevada. Traveling on the Narrow Gauge Railroad, from the Palisades, a station on the Southern Pacific, to Eureka, you can still see the ruins of its plant. It was a sulphide ore that required preliminary roasting and then became tractable and free. Besides the furnaces, the equipment consisted of only a twenty-stamp mill. Yet the ore was of so high grade that the gross production had reached the enormous total of $150,000 in a single month. Much of the ground was totally unexplored, though promising.

Baron Grant laid out his promotion with his consummate skill. He possessed a complete knowledge of the investing public. At that time—and probably still—investors and speculators, as a rule, confined themselves to a single line. One dabbled in coal, another in iron mines, another in silver mines, another in gold mines, and so on down the line. Informed of the specialty of each, the astute baron knew exactly where to go for customers and never wasted time. The plans provided for an issue of £600,000 of common stock and £300,000 of debenture bonds, the latter to be used for a plant to quadruple production.

The enterprise was ably advertised, and this time Samson was tractable and kind. Interest was keen, but I think even Baron Grant was rather surprised at what followed. When the books were opened, there was a crush to get on board, and when we had a chance to assemble figures, everything had been gobbled up and the stock twice oversubscribed. Our net profit was £300,000 or, in American money, $1,500,000.

I had several experiences in the easy-money line, but this put them all in the shade. I was confident that my mission in life was to place American mining securities on the London market. Baron Grant and myself entered into a written agreement. I was to secure options on high-class mining properties. I had in mind the Raymond & Ely, North Bloomfield, Eureka Consolidated, and Zellerbach mines. Grant, on his part, agreed to handle no other mining properties but mine. With this understanding, I did not even wait for the Mineral Hill melon cutting, but set out posthaste for San Francisco to lay in a new stock of options for the foundation of wealth beyond the dreams of avarice.

The news of my success in placing the Mineral Hill mine in London had made quite a stir in my home town, and I was deluged with offers of mining properties, good and bad. Quite a jubilee occurred when the first half-million dollars on account of the purchase price for Mineral Hill was made payable at the Bank of California. The directors of the company were so enthusiastic that they voted themselves $5,000 each as a "souvenir" and added a "souvenir" of $25,000 for the president. The other $500,000 arrived in due season, but the sordid stockholders, who seemed singularly devoid of imagination, objected so strongly to "souvenirs" that this feature of the celebration was overlooked.

I had no difficulty at all in securing options on several of the most assured mining properties of California and the Pacific Slope. From these alone I figured to make millions, judging by the history of Mineral Hill. Figuring on a prolonged stay abroad, I broke up my residence in San Francisco, gave Maurice Dore a power of attorney to manage my local interests, and left with my family for London, to change paper into gold.

CHAPTER XXIV

Baron Grant Demands More Time, Thereby Knocking Out Option for Mine that Soon Developed Bonanza

Exploiter Breaks His Promise and Litigation Follows; Public Fooled into Buying Worthless Securities

I RETURNED to London as soon as my business was arranged in San Francisco. The boom times were still on. Speculation was running mad. I was a trifle chagrined at losing the best property I had bonded through the stubbornness of Baron Grant. This was the famous Raymond and Ely mine. By the payment of $10,000 I had secured an option of this famous property for sixty days for $900,000. I had cabled Grant about Raymond and Ely in order to hasten arrangements, as the time was short. He answered that nothing short of ninety days' option should be considered. I tried to secure an extension, but was turned down. While we were seesawing over this and time was slipping by, the company offered to return my $10,000 and let the option drop. Under the conditions, I accepted the tender. Just a week later the Raymond and Ely bonanza was uncovered, yielding millions in dividends. After that no one could purchase it at any price. Whether the owners really knew anything about this tremendous "strike" when they so generously tendered me the return of my $10,000 deposit, the reader can guess as well as I.

But that was a small matter; I had so many shots in my locker. Among these was the famous North Bloomfield hydraulic mine. I had an option on it for $600,000. In my judgment, which afterwards came true, it was worth at least five times as much. One of the principal owners, Samuel F. Butterworth, followed me to England. Talking of Baron Grant and his power of fascination, I introduced him to Butterworth, who was an able man but cold and unemotional as fate, and after a ten minutes' talk the Baron had him spellbound. "There never was a human being like him," said Butterworth as we retired.

Baron Grant was measurably glad to see me, but not so cordial as the circumstances led me to expect. I spoke to him about the North Bloomfield mine and my desire to have the proposition laid before the public without unnecessary delay, but he seemed singularly backward. At last the cat escaped the bag. He had violated his written contract by agreeing to bring out another mining exploitation, ahead of mine. But my indignation at his absolute lack of faith was nothing compared with my astonishment —almost horror—when he told me that the property he proposed to unload on the British public for a million pounds sterling was none other than the Emma mine.

I had no desire to continue business relations with a man who had shown himself so utterly without faith, but I was at some pains to explain the folly of his project from a mere practical standpoint, setting common honesty to one side. I told Baron Grant that I was familiar with the Emma mine, that Mr. Ralston and myself had recently held an option on the property for $350,000, that I had personally inspected the property and found it a nearly worked-out "kidney," that the principal owner had later offered it to me for $250,000, that I considered it dear at any price. In conclusion, I urged that to promote such a fraudulent concern for a huge sum would not only cause a scandal of far-reaching proportions, but would also ruin the market for American securities for many years to come.

Baron Grant listened coldly. He said he had every confidence in the Emma mine and those behind it. That the proposition had been brought to England by Trenor W. Park, a New York financier, once of California; that it was vouched for by such men of prominence as Senator William M. Stewart of Nevada and by the American Minister, General Robert Schenck. He had no fear of a mine guaranteed by such weighty names. As for my own properties, he said he would take them up when his convenience suited. Otherwise, he possessed the power to prevent any other interest floating them. The interview ended in a violent quarrel.

Even when I demanded my share of the profits in the Mineral Hill deal, Baron Grant held me off with specious promises of speedy settlement, then flatly refused to settle at all. By this time we were sworn enemies. I brought an action for the recovery of £150,000, $750,000 of our money. The Baron harassed me with the usual legal impediments, but in the end, I may say here that I gained a judgment and, what was more important still, collected the amount sued for in full.

I sought new outlets for my mining properties among the highest financial circles of England, not by means of stock-exchange exploitation, but by sales to intelligent and provident investors. The North Bloomfield mine was well and favorably known in England. One of its owners was Tom Bell, an English resident of San Francisco, who cut a large figure in the old days. I had actually arranged the complete details of the sale of this property for £400,000. A meeting was held to confirm the transaction and pay in half the purchase price when an unfortunate remark of Mr. Butterworth caused a halt. He said, doubtless in good faith, that no English manager was capable of handling a California hydraulic mine. But this so offended some of the principal English investors that they quietly withdrew.

In the meanwhile, the Emma mine promotion was

brought out with a brand blare of trumpets. Immense sums were spent in wholesale advertising. The most dazzling and seductive literature was scattered broadcast through the length and breadth of the United Kingdom. Its fabulous wealth was described in the vivid language that fires the speculative spirit latent in every man and in most women. Special stress was laid on the eminent station of the American backers. I have seen much lurid get-rich-quick literature of our own at a time when the industry of plucking the public was unchecked and in full bloom, but nothing that took rank with the effrontery employed to bolster up this brazen fraud.

Of course the promotion was a huge success. When the subscription books were opened, a small river of gold poured in from applicants for shares. The issue was enormously oversubscribed. Baron Grant and his associates selected, as far as possible, the smaller class of investors. These are less able to roar in an effective manner when the inevitable day of reckoning comes for every crooked deal.

The Emma mine was regularly listed on the London stock exchange alongside of reputable and conservative companies. It became the feature of a tremendous gamble. In the hands of expert market manipulators, the stock ebbed and flowed like the tide. Stories of fabulous dividends were passed from mouth to mouth, and the stock soared from one high level to another till ten-pound shares touched £32. This absolutely valueless and exhausted property had a paper value of $16,000,000. When it shrank under profit-taking and selling pressure, reports of new strikes, vast ore bodies uncovered, sent the prices booming once more. Had it not been for the utter heartlessness of the thing, one could almost admire the skill with which a huge deception was organized and kept alive.

Of course I shouted "murder" from the housetops. I publicly denounced the Emma mine as an exhausted, worthless hole in the ground. It was like a voice raised in the wilderness. No one paid the least attention to my warnings in the midst of the bawling crowd. I was classified either as a general calamity howler or as the leader of a "bear" faction, anxious to organize a "bear" raid and interrupt the wave of prosperity. At length, to gain a larger audience and put my statements in responsible form, I made an effort in a new field of endeavor by founding the *London Stock Exchange Review*.

CHAPTER XXV

Inspired by Desire to Expose Emma Mine Swindle, Author Begins Publication of Financial Journal

Ralston Reports Discovery of Immense Diamond Field and Declares His Find Is Worth $50,000,000

Many times I had learned to have a deep respect for printer's ink. I had seen it make history, change fortune, influence the thought of great bodies of people, prove a mighty instrument for good or ill. Without the least desire to be disrespectful to the present, I have a strong impression that the journalism of fifty years ago had a wider dominion over the minds of its readers than the modern school. I cannot say that this was always for the best. Men had a blind devotion to their pet newspapers that amounted to something very much akin to bigotry. Such newspapers were the final authority on everything, from religion to politics, and everyone who questioned their opinions, politics, or statements had a fair prospect for a fight. Thus when an editor fell into some grievous error, he was certain to pull nearly all of his subscribers into the same abyss.

So I realized that to have any influence in my new place of business, to attack the power of Baron Grant, now bitterly antagonistic in every way, and to offset the Emma mine fraud, I must have a personal organ. For this purpose I established a financial weekly paper known as the *London Stock Exchange Review*. It was issued, for apparent reasons, ostensibly by a brokerage firm, but it was an open secret that the publication was mine. I engaged an able editor in chief and directed him to employ the best financial writers in England, giving each his proper department, such as railroad securities, industrial, mining shares, foreign and domestic loans, and the like. I retained a page for which I furnished material hot enough to burn holes in an asbestos blanket. The page was devoted to the Emma mine and Baron Grant.

The *Stock Exchange Review* was a breezy, well-written sheet, full of valuable information to investors, put together in an attractive, readable form. I had set aside £6,000 to pay the losses of the venture. Much to my surprise and pleasure, it proved a money-maker from the start. The style was such a departure from the ordinary dry-as-dust publications of its kind that it made a hit on the street with the first issue. The price was a shilling, but often big premiums were offered when it came out with an extra seasoning of tabasco.

I had reason to know that it tickled the hide of Baron Grant unpleasantly. It managed to hit on a number of raw spots in his past career and in particular interfered with the Emma mine proceedings. While I spoke as a private person, my charges might be disregarded, but when a publication, reasonably responsible in damages and absolutely responsible in a criminal charge, made a downright allegation of fraud, that was quite another thing. The libel laws of England were then, as now, airtight. It was not a jocular affair to call anyone a thief in print, and those who did not seek redress had to suffer under the suspicion, just or unjust, that the accusation was substantially true. The Emma mine was brought out in two sections, for promoting each of which Baron Grant received a commission of £100,000. When the first section was issued, it almost took a squad of policemen to keep back the crowd of investors. The second appeared just after the *Stock Exchange Review* began to hammer. While the stock was taken, the promotion staggered and was never quite itself again. One thing it prevented absolutely—the declaring of a great dividend—on air—which would have sent the stock skyward like a rocket. The managers had determined on this piece of rascality, but doubtless fearing that I had certain knowledge of the fact that not a dollar's worth of precious metal had been produced, this particular piece of villainy was reluctantly abandoned.

In the meanwhile, I began to have suspicions of the integrity of Samson, the financial editor of the London *Times*. I could not think he was ignorant of what was going on in the Emma mine deal. His attitude toward several shady promotions looked, to say the least, queer. In the days of our intimacy Baron Grant had more than once broadly intimated that he possessed some kind of mysterious hold on Samson. Whenever I suggested Samson's name as a possible factor in our enterprises he always said smilingly, "He'll be all right." For some time previous I had a passionate yearning to get a look at the financial writer's bank account.

I communicated this desire to Alfred Rubery one day, just as the expression of a wish. Rubery, who knew all the ropes and all the ins and outs of London, pondered a few moments and said he thought it might be arranged. I was overjoyed at this intimation, but could not exactly see how, though my friend had an odd way of doing things seemingly impossible in an every-day fashion. He was of an impulsive character, a most loyal, trusty and affectionate friend, yet nothing on earth could ever jar him out of his marvelous British self-possession. I remember one occasion in San Francisco in the old *Chapman* days, when Rubery and I were present at a certain meeting, when a wholesale slaughter was on the point of taking place. Just at the crisis I happened to glance at Rubery. He was sitting in his chair with a bored, blasé expression on his face, as if tragedies were so commonplace in his life that they lacked interest and were positively wearisome. This incident has nothing to do with the story further than to illustrate the singular character of the man—a mixture of dash and enthusiasm under strong control.

At that time money could accomplish almost anything in England. A few evenings after our conversation Mr. Rubery waited on me at my rooms, accompanied by a dapper-looking person whom he introduced as an official of the Bank of England who had access to the accounts kept there by Mr. Samson and who was willing, for a reasonable compensation, to give me all the information concerning the same I might desire. To which pleasing presentation the official of the Bank of England gravely bowed.

Some conversation followed as to the terms and conditions. The official wanted all that was coming, but evidently did not wish to scare off a good customer by an extravagant price. At last he got down to a cold cash proposition. Would I consider, under all the circumstances, £500 excessive for so delicate a service?

A bargain was struck readily enough. Much better terms would have been cheerfully granted. The contracting party underestimated his hand.

Through this person I secured an accurate statement of Mr. Samson's transactions. I even was shown checks and cross-checks, which I photographed for future use. These proved beyond all question that Mr. Samson was a beneficiary of the Emma mine promotion, that he had profited largely by other deals, and, in short, was faithless to the trust of his employer and trading on that trust.

Here loomed up the outlines of a great dramatic situation, the unmasking of a conspiracy, the righting of many wrongs, by which the villain of the play would be confounded and the innocent come to their own. Only the details needed rounding out to clear the atmosphere and let the curtain fall.

I was deeply engaged in these great affairs when I received a cable from Mr. Ralston. It was, in fact, a letter. At the cable rates then in force, it cost over $1,100. When I read it, I felt assured that my old friend had gone mad.

It told me of a vast diamond field just discovered in a remote section of the United States. His description of it made Sinbad the Sailor look like a novice. He said that diamonds of incalculable value could be gathered in limitless quantities at nominal expense; that they could be picked up on the anthills; that at a low estimate it was a $50,000,000 proposition; that he and George D. Roberts, a well-known mining man, were in practical control. Finally he almost demanded that I should drop everything, take the next steamer and act as general manager.

The extravagance of his language alone seemed to me to indicate that he was laboring under some strange delusion. However, diamonds or no diamonds, I was in no position to stir. I cabled him briefly that my business in London was of too vital importance to admit of considering other engagements.

But that did not satisfy Ralston at all. Cable followed cable, urging, imploring, beseeching me to come on, which were invariably answered in the same way. Still, I was worried and perplexed. Rumors began to float into London about the discovery of a vast diamond field in the American continent, controlled by the great California banker, W. C. Ralston. Many financiers called on me for information, knowing our relations. Among others, Baron Rothschild sought an interview. He asked me what I knew about the diamond fields, and I frankly showed him Mr. Ralston's cables. He read them with interest and asked me what I thought myself. I told him that while I had great confidence in Mr. Ralston, I thought he must have been imposed upon in some way and that in due season the bubble would burst.

Baron Rothschild mused a moment. "Do not be so sure of that," he said, "America is a very large country. It has furnished the world with many surprises already. Perhaps it may have others in store. At any rate, if you find cause to change your opinion, kindly let me know."

This remark, made by perhaps the keenest financier in the world, was enough to set any one thinking hard.

My position was one of extreme difficulty. The most important engagements of my life demanded my presence in London. Of course I knew that in my absence everything must mark time. But little by little the impression began to grow on me that Ralston had actually captured a fifty-million-dollar financial circus and that I was badly needed as ringmaster. His cables did not deal in hopes, but absolute certainties—assured facts. The diamonds were not a dream—a small fortune of them taken from an insignificant trench were already in his possession. Finally came a cable begging me to go to California if only for the briefest stay, say sixty or ninety days.

I had engaged offices in London for seven years. I could see ahead a vast future of activity and success, and I did not want my selected career broken into by outside distractions, however brief. But I commenced to take the appeals of Mr. Ralston more seriously. Casual expressions of opinion such as the one noted by Baron Rothschild began to stir up my imagination a bit. Could it really be true that there was a place where diamonds could be picked up on anthills? It was very easy to find out the truth, and if the truth happened to correspond with Mr. Ralston's statements, then everything else in the world in the way of business or enterprise seemed commonplace and cheap.

I laid the matter before Alfred Rubery, who usually had a level head. He was surprised at my reluctance. "You have your men safely trapped here," he said. "There is no possibility of escape, and whether they enjoy for a brief time a sense of fancied freedom matters not in the least. Make up your mind to go to California and find out what all this cable correspondence means. Personally, I am bored to death, just pining for a little bit of excitement. I will go along with you, and we will stir up things again in the Far West."

Pressure came on every side. I must have had a forewarning of disaster to have hesitated so long, but finally I gave way to forces that seemed like fate. I cabled Ralston that I would be at his service for a brief period but that the proceedings must be short and sweet. Also I made a hurried arrangement of my affairs in London, thinking to take up the thread again in three months at most. Rubery was rejoiced at my decision and prepared to go along. We turned our backs on London, stayed not on the order of our going when we reached New York, and, as fast as steamships and railroads would carry us, arrived in San Francisco sometime during the month of May, 1872, prepared to uncover the greatest diamond field in the world or return whence we came with equal expedition.

CHAPTER XXVI

Discoverers Decline to Reveal Location of Diamond Field, But Report of Agent Satisfies Promoters

Final Proof of Good Faith Is Offered in Form of Bag Filled With Collection of Eye-dazzling Gems

When I arrived in San Francisco I lost no time in getting in touch with the principals of the diamond deal. Three prominent men only were concerned in it at that time, W. C. Ralston, George D. Roberts and William M. Lent. From them I learned that the alleged discovery of the diamond fields had been known to them for many months. Two prospectors, Philip Arnold and John Slack, were the original locators. I had known Arnold previously in California. He had been employed by Roberts to look into mining properties in the Western country. The later story that he had once been employed by myself in a like capacity was absolutely false. Slack I had known as a plain man about town, of general fair repute.

As an earnest of the great value of the fields, the gentleman had, as near as I can recollect, a large quantity of rough, uncut, brilliant-looking stones which they said local experts had pronounced diamonds of an estimated value of $125,000. Among them were several magnificent reddish-colored stones, said to be rubies. Moreover, they claimed that the discoveries had been verified to an extent sufficient to satisfy themselves.

The story, previous to my arrival, I only know by hearsay, and I cannot vouch for every detail of things beyond my personal experience that happened forty years ago. But as nearly as I can recall the narrative, as it was related, the main facts were these:

One day in the year 1871, when I was in Europe, two weather-beaten men, looking like typical miners, presented themselves at the Bank of California and arranged to deposit property of great value for safekeeping. The property proved to be nothing more than some handsome-looking stones which they said in explanation were diamonds, of which they had discovered a great store in the desert section of the West. They were given a receipt for their valuables and quietly took their leave. But, of course, in those days of mad excitement and crazy speculation, such an incident was bound to leak. George D. Roberts located, in his old prospector Arnold, one of the fortunates and introduced him to Ralston and Lent. Arnold was always the spokesman, the negotiator, in these early transactions. Slack merely was present and acquiesced. At first the men were exceedingly coy and cautious, had all the manner of a couple of simple-minded fellows who had stumbled on something great and, bewildered with their good fortune, were simply afraid to trust anyone with the momentous secret. They declined to give the slightest indication of the locality of the fields or left the impression that they were distant a thousand miles, or thereabouts, from the actual spot. Relying on vague hints, several parties actually set out for Arizona to locate the new Golconda. At the outset the men refused to part with their rights, except to the extent of a small interest, and only then for a large sum of money, which they asserted was necessary to secure claims to a very large territory.

Later, however, they became more amenable to reason. They were willing to part with a half-interest to gentlemen in whom they had such implicit confidence. When it was pointed out to them that negotiations were impossible unless the location of the mines was indicated and some kind of an inspection allowed, they offered a rather strange arrangement, which, however, seemed fair enough on its face. By its terms they agreed to conduct two men, to be selected by Ralston and Roberts, to the diamond fields and allow them to satisfy themselves of the general nature of the find, but with this proviso: that these representatives, after reaching the wild, uninhabitable country, must submit to being blindfolded, both going and coming back. These conditions were agreed to and such an expedition was actually made. I am not certain, but my impression is that David D. Colton was one of the two investigators, being selected by Mr. Ralston as a peculiarly levelheaded man of large experience. However that may be, the mines were certainly visited and displayed, more diamonds were unearthed, and the party returned with the most rose-colored reports of the genuineness of the properties and their fabulous riches. It was this report that set Ralston and his associates wild.

I had some knowledge of the prospectors. Arnold generally had borne a good reputation among the mining fraternity. Slack seemed to be a stray bird who had blown in by chance, probably picked up by Arnold because of a marriage relationship. It seemed that they had told a straight enough story. It was impossible to tangle them in any detail. Still, I had a general, indefinable doubt which I expressed in plain words to Ralston.

Before I arrived, the men made a proposition that seemed eminently fair. This was an offer to go to the diamond fields and bring to San Francisco a couple of million dollars' worth of stones and place them in our possession as a guaranty of good faith. Such a tender was, of course, accepted. Slack and Arnold left San Francisco, promising to be back in record-breaking time.

Shortly after I arrived, Ralston received a telegram from Arnold dated at Reno, stating that he and Slack were on the way and urging that somebody meet them at Lathrop, presumably to share in the heavy burden of responsibility. After a hurried conference I was asked to meet our emissaries as per request, and they were so advised by wire. At the same time a later conference was arranged at my residence. After my marriage in 1866, I had bought the fine family home of Mr. Ralston on Rincon Hill. There my friends were to await my coming till the overland train arrived.

I had a long wait at Lathrop, but at last the expected overland pulled in. I located the men without difficulty. Both were travel stained and weather beaten and had the general appearance of having gone through much hardship and privation. Slack was sound asleep, like a tired-out man. Arnold sat grimly erect like a vigilant old soldier with a rifle by his side, also a bulky looking buckskin package.

Slack soon awoke, and we discussed the business in hand in low tones. The men told a rather lurid story, but yet not improbable in its way. They said they had luckily struck a spot which was enormously rich in stones, which they estimated to be worth two million dollars, that these had been done up for convenience in two packages, one for each, that on the way home they found the water in a river they had to cross extremely high and for purposes of safety had constructed a raft, had nearly been upset, had lost one of the bags of diamonds, but as the other contained at least a million dollars' worth of stones, it ought to be fairly satisfactory.

Slack and Arnold left the train at Oakland, turning over the sack of diamonds on my bare receipt. It was an awkward, burdensome bundle to handle on the ferryboat. Arrived at San Francisco, my carriage was waiting and drove me swiftly to my home. An eager group was assembled.

We did not waste time on ceremonies. A sheet was spread on my billiard table; I cut the elaborate fastenings of the sack and, taking hold of the lower corners, dumped the contents.

It seemed like a dazzling, many-colored cataract of light.

CHAPTER XXVII

Promoters Decide to Submit Samples of Their Collection of Diamonds to Great Authority on Gems

Tiffany Consults His Lapidary and Soon Makes Report that Creates Big Stir in Speculative Circles

I think it was the next day or the day following that a display of diamonds was made in the office of William Willis that filled San Francisco with astonishment. The precious stones were actually displayed in open trays to a multitude of sight-seers until I bought a showcase and gave them some kind of protection. General Dodge, a partner of Lent in mining, bought an interest from the prospectors at once, and Maurice Dore also acquired a small holding, although I do not remember exactly what.

Hitherto there had been no attempt made at organization. It was generally understood that Ralston, Lent, Roberts, and myself owned three-quarters of the properties by virtue of money already advanced and to be advanced. For that there seemed to be ample security in the gems we held. The last invoice alone appeared to be security many times over our cash outlays, to say nothing of the probable value of the diamond fields. So we prepared to get our affairs into business shape without further delay and for that purpose held a meeting at which all concerned were present. The plan of action was to follow these lines: First we were to send a large sample of the diamonds to Tiffany of New York, then, as now, the greatest American authority on precious stones, and have them thoroughly examined and appraised. If their value were proved beyond peradventure, then Messrs. Ralston, Lent, Roberts,

Dodge, and myself were to choose a mining expert to whom Arnold and Slack agreed to exhibit the diamond fields and permit a full examination of the same.

Nothing could possibly be fairer to all, and Arnold and Slack easily consented to these conditions without a moment's hesitation. On the favorable outcome of the valuation and the engineer's report concerning the diamond deposits, we agreed to take care of the financial end. Not connected with any agreement with Arnold or Slack was a plan to facilitate the passage of a law whereby a great territory of mining land could be taken up so as to insure to ourselves the entire field, no matter what the extent. The outline of a corporation was sketched, with a capital stock of $10,000,000 and the allotment of shares to each arranged and defined.

These preliminaries being settled, we set out for New York without delay. In the party were William Lent, General Dodge, Rubery, Arnold, Slack, and myself. It had been arranged beforehand in a general way that Mr. Lent should be president and myself general manager.

We first retained Samuel Barlow, a leader of the New York bar, as general counsel. Mr. Barlow's reputation as a sound business adviser was no less assured than his standing as a great trial lawyer. On his advice we added General B. F. Butler to our legal staff. I had some Southern prejudice against Butler on account of the spoon story in New Orleans, but when I came to know the gentleman, I found him to be very companionable and quite a social genius in his way. A side-reason for employing General Butler was because he was a member of Congress and perhaps able to aid us materially in legislation needed to acquire the diamond fields, as later proved to be the case.

Through Mr. Butler an arrangement was made to meet Mr. Tiffany at the lawyer's house. My counsel had some eye to stage effects. A number of distinguished men were present to see the gems displayed. Among them I remember General George B. McClellan, Horace Greeley, Mr. Duncan of the banking house of Duncan, Sherman & Co., Mr. Tiffany, General B. F. Butler, and the host. I opened the bag of diamonds I had brought from California; also, there were mixed in a few rubies, emeralds, and sapphires. Mr. Tiffany viewed them gravely, sorted them into little heaps, held them up to the light, looking every whit the part of a great connoisseur. "Gentlemen," he said, "these are beyond question precious stones of enormous value. But before I give you the exact appraisement, I must submit them to my lapidary and will report to you further in two days."

Within two days Mr. Tiffany presented his report. In an official statement, still available, his valuation on the lot was $150,000. My own recollection is that he named a much higher sum. However, let it go at that. At that figure we had diamonds enough already in stock to make up a total of $1,500,000 in hard cash whenever we wanted to turn them into money. That certainly seemed a very satisfactory financial basis—regular velvet to begin with. The news of the Tiffany appraisement, though not intended for public consumption, soon became common property in New York and made a big stir in speculative circles. The hardier class of plungers were only too eager to get aboard, even at this early stage of the game.

All that remained now was the choice of a mining expert. One name naturally suggested itself—Henry Janin.

Henry Janin bore at that time in the financial world about the same reputation that John Hays Hammond enjoys today. As a great mine expert and consulting engineer, he was without a peer in the United States, perhaps in the world. Nearly all the big operators like Haggin,

Hayward, and their class were willing to stake their fortunes on his judgment. It was said of Janin that he had the record of having examined something over six hundred mines without once making a mistake, certainly without ever having caused his clients to lose a dollar by his bad judgment. If he had any failing at all, it was on the side of ultraconservatism. Some complained that he never took a chance—that he even turned down good mines to strengthen the confidence of the greatest investing classes, both in the Old World and the New. The O. K. of Henry Janin fixed the reputation of a mining property in every market.

Therefore, there could not have been selected a better equipped expert, so far as the financial world knew, to settle finally the existence of the diamond fields. Mr. Janin was interviewed. He was a man of big affairs whose time was well occupied. But he agreed to make the examination, provided the time to be consumed did not exceed a month. He was also a very high priced professional. His best terms were $2,500 cash, all expenses paid, and a right to take up 1,000 shares of the stock at a nominal price. I may add here that Mr. Janin later on sold his stock, while the excitement was in full bloom, for $40,000. Mr. Lent rebelled and protested against this arrangement as excessive, but was overruled. He and I afterwards purchased Mr. Janin's stock.

At this stage of the proceedings Arnold became restive. He said he was placing his property at the mercy of others without proper security, that what he had received was a trifle compared to the value he was about to disclose, and that he must have a further guarantee in cold cash. He named $100,000 as the amount that must be paid down but agreed to let it remain in escrow, pending Mr. Janin's report. Some quick writing went on between Mr. Lent and Mr. Ralston, as the result of which the latter transmitted the amount by telegraphic order, Mr. Lent holding the diamonds appraised by Tiffany at $150,000 as a further and final security. This was not exactly according to program, but the transaction was fairly businesslike and did not present itself as a holdup.

All our arrangements and differences in New York were settled in a very brief space of time, and we set out in high spirits on the way to the mysterious diamond land. The party consisted of Henry Janin, General George S. Dodge, Alfred Rubery, myself, and Arnold and Slack.

CHAPTER XXVIII

Discoverers of Field of Diamonds Finally Lead the Party of Investors to the Scene of Wonderful Find

Pick Turns Up Many Fine Gems, and Expert Grows Enthusiastic as He Figures Out the Profits

Our friends in San Francisco had, of course, been advised by wire of our transactions and movements, including the date of our departure. We had not journeyed far before we received on the train a telegram that George D. Roberts and a considerable party were on the way to join us to visit the diamond fields and would meet us, if I recollect right, at Omaha. Here again Arnold absolutely rebelled. He had kept all his engagements, he said. The diamonds had been appraised by an expert of our own selection. He was now on his way with the mining engineer chosen by us and with the appointed representatives of the San Francisco interests to exhibit the diamond fields and permit any kind of examination we wished to make, but he was not willing to expose his hand to the whole world until other business arrangements were complete.

It seemed to me that Arnold's argument was unanswerable. Before we left San Francisco, every detail had been arranged. This was a plain departure from the plan. I took sides with Arnold. In fact, there was nothing else to do, for he insisted that everything was off unless we conceded to his wishes. Besides, I had every faith in Janin's ability to pass on the genuineness of the diamond fields. Accordingly, a telegram was sent that turned Roberts and his party back.

We left the Union Pacific Railroad at a small station

near Rawlings Springs. Here we hired the necessary outfit and struck out in the wilderness, Arnold and Slack leading the way. Our course was erratic. At times our leaders seemed to be perplexed, to have lost their way. At times they climbed high peaks, apparently in search of landmarks. The country was wild and inhospitable. We suffered during four days' travel many inconveniences. The party became cross and quarrelsome. At last, on the fourth day, early in the morning, Arnold set out alone, to get his bearings as he said. He returned about noon, said everything was all right, and we set out again with high hopes. By four o'clock we pitched camp on the famous diamond fields.

The spot was at a high elevation, about 7,000 feet above sea level I think. Physically, it embraced a small mesa or rather gently sloping basin, littered here and there with rocks, comprising about thirty or forty acres, through which a small stream of water ran. It was located in one of the most unfrequented parts of the United States, although, as it afterwards proved, Arnold and Slack, in their zigzag course, had actually brought us nearly parallel with the railroad and no more than twenty or twenty-five miles from it. In fact, once, while we were at the mines, on a very still day, I thought I heard something in the far distance that sounded like the ghost of a whistle. When I mentioned this to Arnold, he merely smiled. The railroad was at least a hundred miles away, he said.

But at all events we were mighty glad to reach our destination, and now everything was sidetracked to begin the diamond hunt. We barely unsaddled our animals and secured them, then commenced to hunt diamonds. Arnold and Slack were serene and confident. They pointed out several spots where they had previously dug and found the precious stones already mined and delivered in San

Francisco. We all went to work with our primitive mining implements—picks, shovels, and pans. Everyone wanted to find the first diamond. After a few minutes Rubery gave a yell. He held up something glittering in his hand. It was a diamond, fast enough. Any fool could see that much. Then we began to have all kinds of luck. For more than an hour diamonds were being found in profusion, together with occasional rubies, emeralds, and sapphires. Why a few pearls weren't thrown in for good luck I have never yet been able to tell. Probably it was an oversight.

You may depend upon it that we were in a happy mood that night. There wasn't the usual row over who should cook supper, who should wash the dishes, who should care for the stock, which little incidents of camp life had brought us to the verge of bloodshed during the three previous days. On the contrary, good-will and benevolence were slopping over. Arnold and Slack had excellent reason to be satisfied. Mr. Janin was exultant that his name should be associated with the most momentous discovery of the age, to say nothing of the increased value of his 1,000 shares, while General Dodge, Rubery, and myself experienced the intoxication that comes with sudden accession of boundless wealth.

The next day, prospecting was resumed and covered a wide range. Everywhere we found precious stones—principally diamonds—although a few sparklers of other kinds were interspersed. It was quite wonderful how generally the gems were scattered over a territory about a quarter of a mile square, and of course we were only doing surface examination. No one could tell what depth might produce.

Accounts have been published to the effect that when we arrived at the diamond fields, there were visible evidences of the ground having been tampered with and disturbed. This is absolutely absurd on its face. In the first place any such evidence would have excited the suspicion of the keen-eyed Janin in a moment. Secondly, such a clumsy method of "salting" was unthinkable. Undoubtedly, holes were made in the soil with sharp iron rods, gems were dropped in the holes, which were closed by a hard stamp of the foot, and the first winter's rain obliterated every trace that remained of human agency. Wherever we worked, the ground was "in place."

Two days' work satisfied Janin of the absolute genuineness of the diamond fields. He was wildly enthusiastic. It was useless, he said, to spend more time on that particular piece of property—that was proved. The important thing was to determine how much similar land was in the neighborhood, and be able to seize on everything in sight, for Mr. Janin pointed out that this new field would certainly control the gem market of the world and that the all-essential part of the program was for one great corporation to have absolute control.

So we started on a widely extended prospecting trip. Arnold and Slack did not care to go along, and, to tell the truth, we weren't very anxious for their company. We saw much landscape, also much land that exactly resembled the foundation at the diamond mine. We staked off, in a rough way, an enormous stretch of the country, set up notices of claims that we hoped would hold things down, and covered what we believed to be the entire diamondiferous area.

We returned to the original treasure fields and found Arnold and Slack patiently waiting. Some discussion arose over the vast values we were leaving behind us unguarded and the urgent necessity to place someone in charge. Slack was willing enough to stay, and Dodge and Janin begged me to induce Rubery to remain with him. This Rubery rebelled against lustily. He had come on a pleasure trip—

nothing more. But he was a most accommodating man at heart and finally gave in. So we rode away from the diamond fields, leaving Rubery and Slack on guard. I never saw Slack afterwards—what became of him is a dark mystery that I will take up later on.

CHAPTER XXIX

Public Soon Hears of Wonderful Find and Gossips Carry News Until Whole World Is Keenly Interested

Company to Develop Diamond Fields Includes Great Lords of Finance and One Noted Union General

We returned direct to New York, that is to say, all of the original party except Rubery and Slack. Of course Mr. Ralston was advised by wire of the substantial results of our examination. Likewise, of course, we advised our New York friends who had been previously in our confidence that our best expectations were exceeded. Where so many are cognizant of a secret, it very soon becomes public property, and in a brief space of time all New York and, for the matter of that, all the civilized world knew that vast diamond fields had been discovered on the North American Continent, had been inspected by a mining engineer of great reputation, and pronounced genuine. Something like the profound excitement that stirred the mighty Argonaut movement began to take form everywhere.

As an evidence of this fact, almost immediately after I had reached New York, Baron Rothschild of London, who had previously made inquiries of us, arranged for what amounted to a cable interview. He informed me that he had just received a cable from Mr. Ralston. (This, I presume, related to the agency he accepted at a later date.) He stated further that he had heard of the Tiffany appraisement, also that I had personally made a visit to the mines with a leading expert. He wished me to confirm the result of our observations. I answered Baron Rothschild that half the truth had not been told, that the diamond

BARON ROTHSCHILD

Head of the great financial institution in England in 1872

fields were rich beyond calculation, that every doubt and shadow of a doubt had been absolutely removed so far as I was concerned. The Baron thanked me, saying he was pleased to hear the good news.

In fact, after the Tiffany valuation, the personal examination of the mines, and the statements of Mr. Janin before he promulgated his famous report, every suspicion gave way to an unbounded enthusiasm. Mr. Lent afterward made a written statement, still in existence, that Mr. Janin assured him he could wash out a million dollars' worth of diamonds a month with the assistance of twenty rough laborers. Mr. Janin never went that far with me; in fact, he afterward questioned the entire accuracy of Mr. Lent's figures, and Mr. Lent himself admitted that he might have made an error. But before leaving the diamond field, Mr. Janin assured me that the discovery location alone, which we had partially examined, was certainly worth many million dollars, with countless possibilities besides.

Who wouldn't become enthusiastic with such a showing? It fired the imagination of all financialdom. It upset the caution of the wisest heads in the Old World, as well as in the New. There was a wild scramble to get on board, almost at any price.

Some statements have been made to the effect that I used my influence to have the headquarters of the company at New York instead of San Francisco. There is this much truth to the statement, that it was debated very seriously. This was a plain matter of business, a question of dollars and cents—not patriotism. In launching a concern of such tremendous importance, probably destined to affect profoundly a vast industry, it is always deemed vital to have the support of the largest financial center possible. New York was then, as now, the great haunt of capital in the United States. Many of its leading men were only too anxious to identify themselves with the new exploitation. I did not think that the question of headquarters was one deserving mature consideration, especially inasmuch as San Francisco, controlling the stock issues, would necessarily be the great beneficiary in the long run.

But this point was easily settled. Mr. Ralston and myself owned a majority of the property. This we had held from the outset. I simply telegraphed to Mr. Ralston, laying the matter before him without prejudice. Mr. Ralston's answer was decisive. He said that San Francisco stood ready to furnish any amount of capital required. There was no further argument on that head. To San Francisco the headquarters went, but this much was conceded to New York—that branch offices were to be maintained in that city and that Samuel P. Barlow and General George B. McClellan were to be resident directors, which arrangement was later carried out. The New York connection was clearly indicated by the company's name.

The scene now shifted to San Francisco, where Mr. Ralston had the situation well in hand. A company was regularly organized under the laws of California, entitled the San Francisco and New York Mining and Commercial Company, with a capital stock of $10,000,000, divided into 100,000 shares. Its powers were of the largest possible description, not alone to engage in the business of mining and owning mines and their accessories, but also to engage in every class of commercial business, including the preparation of precious stones for the general market. The apparent intention of the organizers was to move the great lapidary establishments of Amsterdam to the Pacific Coast, and the truth is that this design caused no small concern in the Low Countries, where the cutting of gems is an industry hundreds of years old.

San Francisco was certainly ripe for the new company. Hardly a business man of any considerable wealth would not have considered it a rare privilege to be admitted to participation in the enterprise on the ground floor. It was only a case of choosing the highest class of names in the community to launch the great undertaking under the most brilliant auspices. Twenty-five gentlemen representing the cream of the financial interests of the city of San Francisco, men of national reputation for high-class business standing and personal integrity, were permitted to subscribe for stock to the amount of $80,000 each, and this initial capital of $2,000,000 was immediately paid to the Bank of California.

At a stockholders' meeting the following board of directors were elected to manage the affairs of the corporation: Wm. M. Lent, A. Gansl, Thomas Selby, Milton S. Latham, Louis Sloss, Maurice Dore, W. F. Babcock, William C. Ralston, William Willis. George B. McClellan and Samuel P. Barlow were at the same time elected directors, with headquarters at the city of New York. Mr. Lent was then chosen president, W. C. Ralston treasurer, and William Willis secretary. David D. Colton resigned from his position with the railroad to become general manager.

Only old-timers can recognize what these names meant. All the owners of them are long since dead. Some of them went into a financial eclipse before they died. But in 1872 they stood as the last word in the financial and commercial world of the Pacific Coast. I might mention here for the benefit of the later generation that A. Gansl was the representative of the House of Rothschild on the Pacific Coast.

Such was the line-up. The biggest men of San Francisco were solidly behind the enterprise. Two distinguished citizens of New York represented the company as resident directors there, and in the Old World the famous House of Rothschild became the company's agents. The interest of Slack and Arnold was wiped out finally by a cash payment of $300,000, which was turned over to Arnold personally, he having a properly executed power of attorney to act for Slack. Thus the decks were cleared.

CHAPTER XXX

"Old Miner" Draws on His Imagination and Tells Wild Tale of Single Gem as Big as a Pigeon's Egg

Winter Causes Lull, But Cold Fails to Chill the Ardor of Men Counting on Millions in Spring

On July 30, 1872, the articles of incorporation of the San Francisco and New York Mining and Commercial Company were formally filed, and the report of Expert Janin was made public. As yet, however, the exact location of the diamond fields was undisclosed because the company's rights to the great territories claimed were not completed, although a recent act of Congress changing the mining laws gave ample opportunity. The wildest tales concerning the new discoveries were at once turned loose. An article in the *New York Sun,* signed "Old Miner," located the exact position of the fields somewhere in Southeastern Arizona, a guess that happened to be out of the way by some seven or eight hundred miles. The "Old Miner" further stated that the company had in its possession a single gem larger than a pigeon's egg, of matchless purity of color, worth, at a low estimate, $500,000. You may be sure that this started a good-sized stampede for Arizona.

The directors had several meetings and decided to proceed with extreme circumspection. For one thing they sent a large consignment of diamonds to the House of Rothschild in London for examination and sale. At the same time, a party of fifteen, including miners, surveyors, and others interested, were dispatched to the diamond fields for the purpose of exploring, surveying and securing our rights. In the meantime, not a share of stock was placed on the market, although the excitement was intense.

I append an extract from a morning paper of the day following the incorporation and making public Janin's radiant report:

> (Alta, Aug. 1, 1872.)—American Diamond Fields. One Thousand Diamonds Now in This City. Also Four Pounds of Rubies and Large Sapphires.
>
> We have a wonderful story to tell. We listened to it at first with incredulity, but after hearing all our informant had to say we found reasons for believing it. We have seen a report written by Henry Janin, a mining engineer of an established reputation who had visited the mines, examined them and reported favorably on them. He has accepted the position of superintendent and has expressed the opinion that with twenty-five men he will take out gems worth at least $1,000,000 a month. In this paper he attached so much importance to the discovery that he discusses the question whether the price of rubies and diamonds is likely to depreciate in consequence of increased production and answers the question in the negative. We have thus commenced with Mr. Janin because he is well known here and it is mainly on his statements that confidence rests. His statements evidently command confidence, for some of the leading capitalists of the State have purchased stock.
>
> The place of the new mines has not been communicated to us by any of the interested parties, but street rumor says it is New Mexico. About three years ago, they say, an Indian near the diamond deposits gave several diamonds and rubies to a white man who brought them to Messrs. Roberts and Harpending in San Francisco. These gentlemen satisfied themselves of the value of the gems and sent men to hunt for more. They met the Indian after a long search, he took them to the place and was subsequently drowned. We tell the story as it was told to us.

Then Mr. Janin went to the spot, washed a ton and a half of gravel, took out 1,000 diamonds, four pounds of rubies and a dozen sapphires, and selected the best ground for mining. Three thousand acres were claimed under the mining law passed last session. The country for a considerable distance around was examined, but no equally promising deposit was found.

Most of the diamonds found by Mr. Janin are small, weighing a karat. One obtained previously weighed over 100 karats, but was dark and of little value relatively. There are 109 karats in an avoirdupois ounce, so that a diamond weighing a carat is a small affair, yet if clear and well shaped may be worth from $25 to $50. . . . Some of the saphhires are as large as pigeon eggs.

The diamond mines are the property of the San Francisco and New York Mining and Commercial Company, which has been incorporated, and the directors are M. S. Latham, A. Gansl, W. F. Babcock, Louis Sloss, William M. Lent, T. H. Selby, Maurice Dore, General George B. McClellan and Samuel L. Barlow, the last two of New York. The company is incorporated in this city. It has 100,000 shares of stock and they have been selling at $40, making the present market value of the whole property $4,000,000.

This price indicates great expectations, as Mr. Gansl is the agent of the Rothschilds and Mr. Latham of prominent British capitalists. A party of miners will go to the mines with tools and provisions for the winter's work and the extraction of gems will begin. The stones are to be brought to San Francisco and cut here.

But that was nothing compared with what followed when the last party returned from the fields on October 6. Previous to that, Deacon Fitch had cautioned the public more than once to go slow on the diamond craze. But thereafter even he joined the procession joyously. Witness this:

> (From the *Bulletin,* Oct. 7, 1872.)
>
> The Diamond Fields—About the 20th of August a party of fifteen men left this city to explore the diamond fields about which there has been such a furore of excitement. Amongst them were the following well-known gentlemen: G. D. Roberts, General John W. Bost, M. G. King, M. G. Gillette, Alfred Rubery, John F. Boyd, Dr. C. Cleveland, E. M. Fry, Chauncey Fairfield and Chas. G. Myers.
>
> The members of the expedition returned last evening. They experienced no trouble with the Indians, but had a very tedious march to the fields and thence home. The heads of the party declare that their explorations more than confirmed the original report of Janin of the extent and richness of the deposits and they exhibited specimens which they say they secured with their own exertions with but little labor. This party went merely to explore and prospect the country where the diamonds and rubies were said to abound and not for the purpose of working. They say that active operations could not be carried on when they were there, as the altitude is great and the ground covered with snow. The specimens they brought back are similar to those previously exhibited in this city and they number 286 diamonds of various sizes.
>
> Mr. Roberts says that if they had been deceived they are the worst deceived and cheated men who ever lived. They surveyed 3000 acres of land and propose to keep secret the exact locality until the company receives a Government patent. The implements used by them seem to have been ordinary jackknives—an improvement on the boot heels of the original locators. If so much wealth can be turned up by such primitive means, what might be accomplished with

shovels and pickaxes? The report of the party renewed the excitement and little else is talked about on California street but diamonds and rubies. A meeting of the trustees of the company was called for 2 p. m. and further developments will be awaited with interest. A fact that is so easily demonstrated as the existence of diamonds in that country should not be longer one of doubt and suspicion.

Of course everything was closed down for the winter. But every holder of the company's stock figured on being a millionaire, at least by the early spring, from the proceeds of his diamond field adventures.

I should have added that when we returned from the diamond fields, Mr. Janin took a package of the gems we had found to Tiffany for valuation. We had estimated them to be worth $20,000, but the jeweler scaled this down to $8,000. This didn't disturb Janin. He considered it a "bear" movement.

CHAPTER XXXI

Rude Awakening Follows Dreams of Boundless Wealth; While Promoters Wait for Spring Word Suddenly Comes that They Were Victims of Clever Swindle

Diamond Already Cut Reveals Fraud; Gems Had Been Carried to Scene of "Find" and Planted Like Seeds

Just what might have happened in a single month of wild speculation had the stock of the San Francisco and New York Mining and Commercial Company been placed in any considerable quantity on the market is hard to tell. But one thing is very certain—it would have caused a catastrophe almost without parallel in the civilized world. The public was keyed up to the point of a speculative craze such as even the Comstock never saw, not alone in San Francisco, but in nearly every financial center of the earth. Millions upon millions would have been invested. The shares would have soared to fabulous figures. Banks would have advanced money on these prime securities, as was the custom in those times. And then the awful crash! There would have been more ruins in financialdom than San Francisco exhibited after the fire. Every day the mails were loaded with letters from eager correspondents making inquiries for stock. The best and unanswerable proof that everyone connected with the company acted in absolute good faith is to be found in the fact that not a share changed hands.

Meanwhile, however, handsome offices were engaged, and David D. Colton was installed in all the dignity of general manager. This was long before the date of typewriters, and it required several clerks to answer letters.

A large map showing the general outlines and physical characteristics of the 3,000 acres claimed by the company was displayed in the office. It showed the relative position of Discovery Claim, Ruby Gulch, Diamond Flat, Sapphire Hollow, and other locations with names equally suggestive of wealth without limit. Many longing eyes were cast on that map by would-be speculators. The company had considered a plan for holding and working what was known as Discovery Claim on its own account and granting concessions in the remaining territory for so much down in cash and a royalty on the gems recovered.

Some fifteen or more bona fide offers were made to purchase a concession for $200,000 cash and a royalty to the parent company of 20 per cent. Not only that, but the purchasers of such concession would have been able to place stock on the market and sell the shares like hot cakes. Quite a few million could have been gathered in from that source alone. Why not? Even granting that the element of gambling was strong, nevertheless, such a property had a far better backing of apparent value than nine-tenths of the wildcat mining schemes launched every week on the stock exchange.

Not only that, but three other diamond and ruby companies were organized, each with fairly representative men behind them. One of these companies exposed to public view a gem that looked like the headlight of a locomotive seen through a fog after dark. It was known as the Staunton Ruby and was generally conceded by experts to be a genuine stone of high quality. No one seemed able to give more than a guess at its value, but the opinion was unanimous that only some rich and powerful nation could purchase it to adorn a scepter or a crown. All of these companies were merely marking time, waiting till the great, proved, unquestioned company should say "play ball" and start a speculative market for everyone.

But no such misfortune happened. On November 11, a telegram was received from Clarence King by the president of the San Francisco and New York Mining and Commercial Company, dated from a small station in Wyoming, stating that the diamond fields were fraudulent and plainly "salted." This, of course, caused a wild excitement among the officers of the company. They held a hurried meeting. They were simply stunned. King was reached by wire at once and agreed to take a party in and prove his statements. A party was at once organized for this purpose. The members were Henry Janin, D. D. Colton, John W. Bost, and E. M. Fry.

Clarence King was a geologist and engineer in the service of the United States government, a man of some professional distinction and talent in the literary line. It is worthy of note here that some years after the diamond story broke, King wrote a perfervid narrative entitled "Mountaineering in the High Sierras." In it he described an ascent of Mount Whitney, the highest peak in California, and dragged himself through a series of hairbreadth escapes that put every Alpine adventure in the shade. A geologist by the name of W. A. Goodyear knew something of the region, visited Mount Whitney, made the ascent on a mule with settled habits of reflection, and never dismounted till he reached the top, proving that King had never been there at all. All of this Goodyear described in a widely-circulated magazine. The laugh that followed broke King's heart. He died a few months later.

It was this same gentleman who, late in the fall of 1872, made up his mind to have a look at the diamond fields. Notwithstanding all of our attempted secrecy, almost anyone could place his finger on our claims. Not only that, but at least two men, Berry and McClellan, had actually been at the fields, saw the old washings and the tools left by the Roberts party, and it was one of those who guided Mr. King to the spot.

Mr. King's story makes the discovery of the fraud rather a matter of deductive reasoning, whereby little straws of evidence are put together one by one and formed into the nest that holds the egg of proof. It is easier to construct this nest afterwards than before. I heard myself a somewhat different version of the story. In company with Mr. King went a middle-aged German, a sort of cross between a camp follower and a friend. Like a "super" in a great dramatic performance, he did not cut a very large figure. But many years afterward I met him in New York, and he told me a very interesting story. On reaching the diamond fields, he said, notwithstanding the intense cold weather, both he and Mr. King began washing for diamonds and, naturally enough, found what they were looking for. In fact, the geologist came very near being fooled as badly as anyone else—wanted to leave instantly and thought of going to San Francisco to have a talk with the directors of the company. But the German gentleman felt differently. He was not overburdened with wealth, had never been in any place before where diamonds could be picked up without even saying, "by your leave," and he was naturally averse to leaving a place so full of delightful possibilities. So he arranged a brief respite before departure. In the meantime, he was washing "dirt" to beat the band and every now and then pocketing a sparkler that he valued at a small fortune. Suddenly he came across a stone that caught his eye and filled him with wonderment. It bore the plain marks of the lapidary's art. He took it immediately to his principal. "Look here, Mr. King," he said.

"This is the bulliest diamond field as never vas. It not only produces diamonds, but cuts them moreover also."

King grabbed the half-cut diamond. Everything was clear as day. Beyond the peradventure of a doubt, the fields were salted. He hunted out evidence that he had overlooked before and very soon was in possession of proof, quite aside from the partly cut gem, that a wholesale fraud had been committed.

I am not giving this story as a fact—simply offering it for what it is worth, and certainly without any desire to detract from the great service rendered by Clarence King.

Mr. King reached the diamond fields on November 2, 1872. On November 10, he was back to the railroad and sent the famous dispatch—that the company was duped.

Also, he waited for Messrs. Janin, Colton, Bost, and Fry, the party sent from California. They went together to the diamond fields, and now the plain nature of the plot was thoroughly exposed. It is not necessary to go into any of Mr. King's geological conclusions or the entire evidence upon which the conclusion was reached. Two or three facts are enough to indicate the satisfactory nature of the proof.

Mention has been made of anthills sparkling with minute but veritable diamond and ruby dust. Perhaps because they were so pretty no one ever disturbed them. But if somebody had taken a notion to give one of them a kick, their supposititious nature would have been apparent. They weren't anthills at all. They were fakes, the work of a sinful man, not of the moral insect. They were also works of art; no one would have suspected guile from looking at them.

A close examination revealed three holes evidently made with a stick or some sharp instrument, at the bottom of each of which a gem rested. There is little doubt that all the "salting" was done in this way, except that as a rule the holes were carefully closed. But in such extensive operations a little reckless work was likely to slip in.

Finally, on the top of a large flat rock, several rubies and diamonds were found pressed into crevices to hold them in place. This was so grotesquely raw that it seems incredible, and led to a story that some of the diamonds were in the forks of trees. Unfortunately for the story, there weren't any trees in the neighborhood.

The party returned to San Francisco late in November. On the twenty-fifth of that month the general facts were given to the press, that the diamond fields were a fraud, and that everyone had been taken in. The excitement was intense. The Associated Press kept the wires humming with the news for days, transmitting fuller reports than were published here, although the local papers printed whole pages. Wherever a printing press ran, the world knew the story of the diamond fraud.

The trustees of the San Francisco and New York Mining and Commercial Company held various meetings, and a select investigating committee was appointed. W. H. L. Barnes was the company's regular attorney. Messrs. Hall McAllister and S. M. Wilson were added to the staff to ferret out and punish those guilty of the fraud. Everyone connected with the early history of the transaction gave testimony; every line of evidence was hunted down.

Among other things, an accomplice came forward by the name of Cooper, who admitted with noble candor that he was the author of the whole scheme, though unrighteously deprived of his welching partners and of his just share of the spoils. Salting mines was the commonest thing in the past and isn't yet to be classed with the lost arts. Talking with Arnold and Clark [Slack?], whom he knew personally, of how the "salting" of gold and silver mines had been overworked, he suggested the "salting" of a diamond field as a pleasing variation, and told how small diamonds, such as those used for drills, could be readily obtained. According to his story, Arnold and Slack bit greedily, and a triumvirate was formed to carry on the fraud. This was nearly two years before the Janin examination. Cooper was undoubtedly a confederate, did a lot of advising and suggesting, but was kept in the dark concerning the most important details. Also, he was promised a liberal share in the booty, and his confession was prompted chiefly by a desire for revenge. He gave Arnold and Slack the full credit for everything.

The statement of Cooper was made not only to the special investigating committee, but also to the grand jury of San Francisco. The latter body indicted no one.

On November 27, the trustees of the San Francisco and New York Mining and Commercial Company met for the last time. At this session it made a final report to the public, giving its brief history, the confidence placed in Tiffany appraisement, and the report of Janin, the final statement that the properties it claimed to be diamondiferous were "salted," and that everyone had been cleverly duped. All its business was summarily suspended and its attorneys ordered to wind up its business at once.

Appended to the report were statements from Clarence King, describing his discoveries, from Henry Janin, confirming Clarence King and admitting his former errors, also from Messrs. Colton, Fry, and Bost, all denouncing the fraud.

If anything were lacking, news came from London that the diamonds we had sent there were coarse, almost valueless "niggerheads" from the South African fields and had been purchased in bulk there from a dealer nearly a year before, who identified them perfectly.

The late diamond millionaires, who had been rather chesty, presented a sad spectacle on the street. They were pursued everywhere with jibes and jokes. Some of them went into retirement till the storm blew over. There never was a better illustration of the joy to be found in triumphing over the sorrow and discomfiture of others.

CHAPTER XXXII

Victim of Big Swindle Explains How Rough Miners Managed to Deceive Men Like Tiffany and Janin

Inquiry Reveals that "Salting" of Diamond Field Cost Plotters $35,000 and Yielded $600,000 Net Profit

How so many of the shrewdest men in the world could have been absolutely duped by the great diamond fraud may well be asked. The truth is, it succeeded not because of the baleful craft employed in working out its details, but because of a rawness that seemed to disarm, rather than arouse suspicion, and the audacity and nerve with which everything was carried out. That diamonds, rubies, emeralds, and sapphires were found associated together—gems found elsewhere in the world under widely different geological conditions—was a fact that ought to have made a goat do some responsible thinking. But it seems to have been entirely overlooked by Tiffany, by Janin, by the House of Rothschild, to say nothing of Ralston, Sam Barlow, General McClellan, General Butler, William M. Lent, General Dodge, the twenty-five hard-headed business men of San Francisco who cheerfully invested $2,000,000 in the stock, and the fifteen mining men who accompanied Mr. Roberts to the fields after the San Francisco and New York Mining and Commercial Company was organized. Had serious attention ever been directed to that single point, it certainly would have prompted an investigation that must have ended in exposure.

Again, the Tiffany appraisement of $150,000 on not more than a tenth of the gems actually on hand is hard to comprehend unless regarded in connection with another fact—that valuing cut stones and valuing stones in the rough are widely different matters. While the Tiffany establishment had undoubted experts as to the finished diamond, it is doubtful whether a single real expert valuer of rough diamonds was to be found in the United States. All the lapidary work of the world was then done principally in Amsterdam, with smaller establishments in Paris and London. As I understood later, the Tiffany experts satisfied themselves that the stones were actually diamonds, weighed them, estimated the cost of cutting and net weight, applied the usual rules for valuation, which increases enormously with the size of the stone, made a large deduction, and let it go at that. Thus the total of $150,000 was arrived at. Knowing the immense reserves we had in San Francisco, the question of fraud probably never entered their minds. For, although much ingenuity and some money had been invested in the enterprise of palming off worthless mining properties, it seemed the height of absurdity to suppose that anyone had invested $1,500,000 in "salt."

And, to do justice to Henry Janin, I think it was this valuation that disarmed his suspicions and made him less eager to search for traces of chicane. He said several times on the journey to the diamond fields that he considered their genuine character established, that his mission was mainly to estimate their extent and probable value. As to that, the washings we made might well have satisfied any man. Perhaps had he remained at the discovery claim instead of exploring the country in the neighborhood, he might have detected traces of fraud. But he considered the most essential thing was an examination to determine the diamondiferous area, so that his employers might ultimately get it all.

Yet the most convincing factor of all was the attitude of the men themselves. Arnold was no ordinary mortal.

Throughout all the negotiations, coming in contact with some of the most alert intellects of the time, he was always serene, ready, confident—did not make a single break. Besides, he had an air of simple, rugged honesty that impressed everyone he met. General Dodge, who thought meanly of human nature, said in a printed interview that he would stake his life on Arnold's integrity. Not only that, but Arnold and Slack were willing, even eager, to submit the diamonds to any test and to lead a party of experts to the fields, under proper guaranty that their rights would be protected. They seemed almost exultant when they understood that Tiffany would value the diamonds. That, of course, would settle everything, they said. They were equally delighted at the choice of Janin as an expert. Both of them had the dramatic gift highly developed. On the stage they might have made the most famous actors of any time.

And how did a couple of ordinary prospectors secure the very large sum undoubtedly used to finance the glittering fraud? That was a question that puzzled many and led to all kinds of surmises about confederates, syndicates, and so forth. But it was shown later that in 1870, Arnold and Slack made a couple of lucky turns at selling mines and actually had at one time in excess of $50,000 to their credit in a Western bank. This deposit was withdrawn in bulk and was never traced afterwards, except in the purchase of diamonds in the markets of Amsterdam and London.

Through the agency of I. W. Lees, this end of the transaction was fully traced and the facts published. Arnold made two trips to Europe to purchase gems. Both times he shrewdly avoided American ports, sailing and returning by way of Halifax. His first visit was made in the fall of 1870. That time he confined his activities to Amsterdam and showed great shrewdness in concealing his tracks and avoiding suspicion. He visited the various gem-cutting establishments, bought many coarse stones, but not enough from any one firm to make the transaction look unusual. No one seemed to know his name, but his photograph was at once identified by many diamond dealers of Amsterdam as the eccentric person who seemed to have an unusual penchant for inferior stones. He was regarded as a newly rich American, with a vulgar taste for ostentation, who wished to overburden his family and dazzle his fellow countrymen with a wealth of cheap, almost worthless gems.

On his second trip in the early winter of 1872, Arnold went direct to London, and there, while the conspiracy was at the most ticklish point, he threw all caution to the winds. One of the largest dealers in the great metropolis gave the story to the press how one afternoon a rather rough-looking American appeared at his place of business and asked to be shown what they had in the way of undergrade, or rather refuse diamonds. He was shown a large stock of South American stones of the quality known as "niggerheads," handsome enough but of very small commercial value. The American pawed over them, apparently without the least regard for size or quality, until he had collected a great pile. Then he asked indifferently, "How much for the lot?"

The trader hadn't the least conception that his customer meant business. However, he made a rapid appraisement of the stones and gave the price at £3,000, or $15,000. To his amazement the American produced a huge bank roll, counted off the money, had the diamonds packed in small sacks which he deposited in the capacious pockets of an overcoat and elsewhere, said good day, and departed. In the photograph of Arnold, the English trader recognized his customer at once.

As near as anyone could estimate, about $35,000 was invested in "salting" the claims. To this should be added something for traveling expenses, etc. The men received approximately $660,000. That left a little over $600,000 net profit.

CHAPTER XXXIII

Principal in Diamond Swindle Goes Back to His Old Home in Kentucky to Enjoy Hard-Earned Riches

Victims Bring Suit for $350,000, But Arnold is Popular with Neighbors and Forces Compromise

AFTER ARNOLD received his final payment of $300,000, he retired to his old home at Elizabethtown in Hardin County, Kentucky, bought a fine piece of land and also a safe, which he kept in his house under strong guard. In this he deposited nearly all his spoils, although he also had a tidy balance in the local bank, which added greatly to his repute among his neighbors. He had a host of relatives in Hardin County, which borders on the primitive section of Kentucky. It was there that the most capable of Morgan's guerrillas were recruited, and there most of them returned. Anyone hunting trouble in that locality was almost sure to find it. Arnold settled down quietly among his friends and relatives to enjoy the fruits of a toilsome life.

His place of residence was well known. In fact, the Kentucky papers gave some prominence to the return of this famous discoverer of diamond fields to the home of his ancestors. When the bubble burst, Mr. Lent hurried to Kentucky, hired eminent counsel—Judge Harlan, later a justice of the Supreme Court of the United States, and Benjamin Bristow, a lawyer of equal standing—brought suit against Arnold for $350,000 on his personal account, and levied an attachment on his property. All of these proceedings are set forth in the *Louisville Journal* of December 18, 1872. Two days later, the same paper published a long statement from Arnold in which he denounced in unmeasured terms the outrage that had been committed on his rights. He scored "Bill" Lent in language of scant courtesy, but of picturesque Western expressiveness, and declared he neither owed him $350,000 nor the like number of cents, or any other sum for the matter of that.

Arnold went on to say that his safe contained $550,000, the result of arduous labor as a prospector and miner in the Far West, not to mention his bank account and real estate. The sequestration of the same by a shark or an aggregation of sharks from California he looked upon as an outrage unparalleled in history. He went into the diamond field story in detail, denied that he had ever "salted" it or that it had ever been "salted" at all. He appended Janin's report, the Tiffany appraisement and a long extract from the *San Francisco Chronicle* to prove that he had turned over an absolutely valid diamond property to the San Francisco and New York Mining and Commercial Company, and that if anyone "salted" it, the diabolical act must have been done after the experts' examination and by some of the "California scamps."

Did Arnold suffer any in the estimation of his compatriots by reason of the grave accusations preferred against him? Rather the reverse. They gloried in what they were pleased to call his "spunk." The old Morgan raiders and thousands of their way of thinking looked with pride, almost with reverence, on one of their kind with nerve and wit enough to make a foray into Yankeedom and bring away more than half a million in spoils. To tell the truth, Arnold was the very hero of the hour, for the old war feeling was still rampant.

I followed Lent to Kentucky, whither also went Captain I. W. Lees. Familiar with the field, after some investigation of the state of public opinion in Hardin County, I am satisfied that had Arnold stood his ground unflinchingly not a dollar could have been wrung from him by legal proceedings, no matter what the proof. And, moreover, at that time the matter of exact proof was not as easy as later on.

Negotiations leading to a compromise took place in which I played a part. These resulted in a compromise by which Arnold surrendered $150,000 on consideration of immunity from further litigation. The money was turned over to Mr. Lent personally. What disposition was made of it I am not informed, but [I] understood that it was retained by the recipient to make good his personal loss.

So Arnold, left, according to his own statements, with an uncontested fortune of nearly half a million dollars, everywhere enjoyed the esteem and high respect which broadcloth and a large cash balance invariably inspire. But he did not live long to enjoy his prosperity. Arnold, among other ambitions, wanted to shine in finance and for this purpose opened a bank in Elizabethtown and for a time did a rushing trade, to the great irritation of his business rivals. The quarrel became very bitter, and as differences of opinion were only arbitrated in one way in Hardin County at the period mentioned, the first time Arnold met one of his competitors, the two opened fire at each other on the street after the manner of the best traditions. Arnold never lacked courage and had all the best of the arbitration, having winged his man once, when his antagonist's partner appeared in a doorway and landed the greater portion of a charge of buckshot in the diamond discoverer's shoulder. His wounds were considered fatal, but his iron constitution carried him far toward recovery, and he was considering with pleasant anticipation a second meeting with the bankers, with sixshooters instead of a clearinghouse to balance the account, when he was seized with pneumonia. Under this last affliction, the tough old campaigner, after a hard struggle, weakened and died. This happened, I think, near the close of 1873, so that Arnold's prosperity was short lived.

What became of Slack? That was a question often asked but never answered in a satisfactory way. As I said, the last time I ever saw him was when I left the diamond fields with the Janin party. He and Rubery remained behind. When these two separated, Rubery came to San Francisco, while Slack took an eastbound train. Many attempts were made to locate him at a later day. He was heard from at various points—St. Louis, New Orleans, Memphis, and Mobile. Always it turned out to be another Slack. Finally the impression became general that he must have gone abroad and hid his identity in another land.

But the strange part of it was that Arnold had all the money, or nearly all of it, as appears by his signed statements and later by the inventory of his estate, which corresponded. Granting every possible contingency, the share of Slack was either practically nothing or very small, not to exceed $30,000 at the utmost. As they always figured as partners and as Slack, though not the spokesman, appeared a man of force, I have always considered that a deep mystery hung over his fate. It seems not unlikely that he died somewhere in the Western country, probably among strangers, and never participated in the profits of the diamond fraud at all.

CHAPTER XXXIV

Diamond Fraud Loss Falls on Shoulders of Original Dupes; Ralston Reimburses All Stockholders

Gossips Make Unjust Charge Against Men Who Acted in Good Faith and Were Deceived by Swindlers

The losses growing out of the diamond fraud fell on the shoulders of the original dupes—W. C. Ralston, William M. Lent, George Dodge, and myself. My impression is that the money obtained by Mr. Lent from Arnold very nearly, if not quite, balanced his account. Perhaps he may have given a portion of this to General Dodge, his business associate. Mr. Ralston promptly paid the twenty-five stockholders, who subscribed $2,000,000 for a half-interest in the company, dollar for dollar. Not a man of them lost a cent. This involved a sacrifice of the last $300,000 paid to Arnold and Slack. Mr. Ralston had the receipts in full of the various parties neatly framed, and I am told that it was one of the mural decorations of his private office in the Bank of California. The remaining balance of loss was borne by Mr. Roberts and myself.

The diamond-fraud story has covered acres of newspaper space. This, however, is the first time that the narrative has been told from start to finish, all the facts assembled in connection form. From what has gone before, certain points stand out in bold relief.

The scheme, or rather the execution of the scheme, was anything but the work of a farseeing, skillful, and well-informed mind. Nothing can illustrate this better than the supreme folly of planting diamonds, rubies, emeralds, and sapphires in the same matrix. A capable rogue would have consulted the history of mining for precious stones and would have readily discovered that they are never found associated in the same formation. This would have enabled him to avoid a raw monstrosity that should have led to exposure at the very start. Much of the other work was raw, as, for instance, the diamond- and ruby-spangled anthills and the flat rock whose fissures were studded with precious stones. A plain, unornamented diamond field would have presented a far better baited hook.

It can be shown by authenticated documentary evidence that Slack and Arnold were the sole beneficiaries of the loot. In fact, some doubt exists whether Slack was a participant at all. Outside the sum that Lent collected, the balance was transmitted to Arnold's heirs, as the records of Hardin County, Kentucky, prove.

Again, if any other actor in the drama had the least foreknowledge of the fraud, he surely would have parted with his interest while the market was booming—before the frail bubble burst. When the coarse diamonds were sent to London, nothing could be more certain than their immediate identification as nearly valueless South African stones. Yet not a share of stock was sold. Every reasonable presumption pointed to the entire good faith of all, so far as the San Francisco and New York Mining and Commercial Company and its stockholders were concerned.

Finally, it is conceivable only on the basis of downright madness that any man with wealth, reputation, and self-respect, in short, with everything to lose, could have conceived and carried out such a reckless plot. If by any chance it had succeeded, if the diamond company's stock had been exploited on the stock market, there was not a place upon the earth so desolate and remote but that the vengeance of mankind would have found him out. It was the evident design of a rather crude intelligence, utterly regardless of consequences and counting on obscurity to make good.

Nevertheless, no matter how plain a case may seem, no matter how free from doubt or complication, if it be only big enough, the world loves to build around it a fairy structure of mystery or romance. Nothing could be more evident than that Arnold and Slack were the architects of their own work. Yet the public saw fit to cast grave suspicion on those who were clearly victims and heavy losers—the only ones who lost a cent. Even Ralston, although he paid over $300,000 to make good the losses of the stockholders, was more or less under a cloud. Lent, Roberts, Dodge, and myself were in turn suspected. At last public opinion seemed to settle down to a conviction that the guiding—"the master mind"—was mine.

There was not an atom of valid evidence on which to raise the accusation. On this side of the Atlantic it was never remotely charged in any responsible paper, to my knowledge. The conclusion seemed to be reached very much because of the largeness of my business undertaking and my well-known spirit of venture in commercial lines. Therefore, not a few assumed that because I was always willing to take what might be called by some a long chance, therefore, I must have been the power behind the scenes with Arnold and Slack.

Not arguing the case, this conclusion had to put aside many well-known facts, as I said before. I was a man of large wealth, making money as rapidly as was good for anyone, so that the financial inducement was not there. I was young, only thirty-two, had a family of which I was proud, had the best possible standing with businessmen, both in San Francisco and abroad. Honor bright, does it not seem incredible that a man situated like myself, full of ambition and with everything to live for, would have engaged in an ignoble plot to fleece his friends and the public, a plot absolutely certain to drag him and all belonging to him through the dust?

The story would have died a natural death, beyond any question, just as it did in the case of my fellow victims had not the London *Times* made a direct accusation of complicity in the diamond fraud against Alfred Rubery and myself which became the basis of a famous libel suit.

CHAPTER XXXV

BARON GRANT BOBS UP AGAIN; TRIES TO GET EVEN ON MAN WHO EXPOSED ONE OF HIS BIG STOCK SWINDLES

Alfred Rubery Brings Suit Against London Times *for Libel and Is Awarded £10,000 as Damages*

IN THE CHARGES made by the London *Times* it was not difficult to recognize the handiwork of my old enemies, Baron Grant and the financial editor, Samson. The accusation seemed to be an echo of the old Emma Mine fight, when I warned the public against the exploitation of a worthless property. That bubble had burst, carrying ruin to investors, disgrace to the promoters, and more than a decade of distrust for every American security in European markets. But the sting of defeat remained, and the opportunity to retaliate was one not to be overlooked.

Alfred Rubery, being a British subject in good standing, brought the libel suit against the London *Times*. As my intimate and close companion for nine months, covering the various incidents involved, he admitted that whatever involved me involved himself as well. Although the earth was ransacked for evidence to connect us with the fraud, the defense absolutely failed to sustain the newspaper's charges. Not only that, but the proof I had gathered, as described in a previous chapter showing the secret bond between Baron Grant and the financial writer, was thoroughly exposed, ending in the ruin of both. Samson was dismissed in disgrace by the London *Times*. Enough was shown of Baron Grant's methods to involve him in lawsuits innumerable that stripped him of his fortune in the end. He did business under assumed names long after, but never with his old success.

Heavy damages were awarded Rubery—the sum, if I remember right, was £10,000. Years afterwards he moved to Australia, and as I never heard from him after, I presume that like the other actors in the diamond-field drama, he is dead. In fact, of all who were in any material way connected with the historic incident—and there were many—I alone survive.

For myself, I felt crushed beneath the burden of vague suspicion, became disgusted with life in general and with business in particular, and formed a determination to retire permanently from active affairs at once. With this end in view I offered my extensive California holdings on a dead market and accepted bargain prices. My controlling interest in the Montgomery Street Land Company I sold to Messrs. Ralston and Sharon, so that they owned share and share alike. I sold a great acreage of tule land to George D. Roberts, part of which comprises what is known as Roberts Island, not far from the city of Stockton. A large estate around Honey Lake I disposed of to various purchasers. Scattering investments in San Francisco were cleaned up in a summary way. I would hardly care to know what all these properties are worth today.

In four months after the diamond fraud was exposed, I had converted into cash everything tangible I possessed on the Pacific Coast. Although the sacrifice I made was enormous, I realized more than $1,250,000, which was as good or better than $5,000,000 today—a fortune ample to supply the most extensive and up-to-date wants of modern times.

The great mistake of my career, entirely apart from monetary reasons, was this hastily taken resolution to seek the shades of private life. Had I faced the music like all the rest—like Ralston, Lent, Roberts, Dodge, and one or two other original "dupes," I would have outlived every trace of suspicion, just as they did themselves. And I am glad to give evidence at this late date, long after all of them are dead, that they were as innocent as children throughout the whole transaction—were the unhappy victims of a costly confidence in men.

But as I took a pessimistic view of things in general and saw fit to withdraw from public view, perhaps I have not so much reason to complain because, in my absence from the world, Dame Rumor was busy with my name.

Three alleged histories of San Francisco, which profess to give an accurate narrative of events, devote much space to the diamond-field fraud. Considering the mass of documentary evidence easily accessible, the misstatements of many facts and the omission of others is noteworthy and may call into question the entire accuracy of all these works. To go no further, they all agree that the losses of stockholders were enormous, claim that they brought suit in the state of New Jersey against Arnold and Slack but never recovered a cent. Lent's suit in Kentucky, the only place where such an action could be maintained, is not mentioned, nor the $300,000 which Ralston contributed to make good. Under these conditions I should not feel hurt because they surmise that the plot was conceived in the "active brain of Asbury Harpending."

I returned to Kentucky, made considerable investments in agricultural land, and settled down to play the part of the country gentleman. My estate was one of the finest in southwestern Kentucky and became a center of hospitality in its region. And there I made another grave mistake—not to remain content with the finest existence in the world, that of an independent owner and tiller of the soil.

Baron Grant Bobs Up Again

It was while I was living in my new home in Kentucky, at peace with all mankind and oblivious of the outside world, that I had a sharp and vivid reminder of the unforgotten past when the papers told, one day in August, 1875, of the failure of the Bank of California and two days later the tragic story of my old friend Ralston's death.

CHAPTER XXXVI

Associates Bar Great Financier from Conference and Soon After His Body Is Found in the Bay

Fortune Plays Cruel Trick; At Height of Ralston's Power His Big Bank Is Forced to Close Its Doors

Ralston succeeded D. O. Mills as president of the Bank of California, in 1872. While conceding the titular supremacy to another and contenting himself with the station of cashier, Ralston had always been the actual head. In all matters of policy and large accommodation his word was law. After the withdrawal of Mills the directors practically gave him a free hand.

All through the ascendancy of Ralston the institution had the splendid reputation that the Bank of California enjoys today. It not only possessed the fullest confidence of the community, but ranked as one of the strongest banks of the United States, with agencies throughout the civilized world and unlimited credit everywhere. The splendor of Ralston's hospitality, the immense enterprises in which he was engaged, and his vast holdings in real estate and corporate concerns gave him the standing of a man whose wealth was almost beyond computation. There was not an intimation of embarrassment when, on August 25, 1875, the Bank of California closed its doors.

I was not a witness of what followed. I was living quietly in my home in Kentucky. But from what I have heard, it was one of the most intense moments in the history of the West. For blocks around the Bank of California stood a packed mass of pale-faced men, anticipating ruin. What might have become an unparalleled panic and almost universal wreck was happily averted by the closing of the stock exchange and the practical suspension of business for a period long enough to allow the community to catch its breath.

Ralston stood the ordeal with all the resources of fortitude, met many patrons of the bank, admitted the grave conditions of its finances, but contended that its assets were very large. Everything he possessed in the world, he said, would be used to make good.

On the afternoon of August 27, the directors called a meeting. Ralston was on hand as usual, but was barred from attendance by D. O. Mills. The incident, it is said, touched him to the quick. Every director had profited by his friendship in the days of his power and prosperity. This seemed a harsh return in the hour of his deep distress. He left the meeting with a dazed and haggard face.

He proceeded to his home and thence to North Beach, where he was accustomed to take a swim in the bay when the weather was opportune. There are a number of living witnesses of what followed. Shortly after he entered the water, other swimmers noticed that something was amiss. His body did not sink, but he was floating face downward. A boatman was quickly at his side. The boatman declared that the banker was still living. Be that as it may, when he reached the shore with his burden, the once master spirit of the Pacific Coast was dead.

How great was the hold that Ralston had on the hearts and minds of men could only be illustrated by the passion of grief under which the whole city bent. His death was looked on as a common calamity. No spectacle has ever been witnessed in modern times such as his funeral presented. By common consent, business of every kind was suspended in San Francisco. You might say that the population of the city of more than 150,000 inhabitants turned out en masse. The proudest and humblest touched shoulders at his grave. Such tribute was never paid to any potentate or prince.

There were even some who found it convenient to make an exhibition of immoderate sorrow who might have been more fittingly employed elsewhere.

I once heard a story of a French gentleman who had suffered a domestic bereavement. A friend met him shortly after and tendered the customary condolences.

"Ah! My poor wife! Yes, it was indeed a great loss!" sighed the Frenchman.

"I was at your house during the funeral," continued the sympathetic friend, "and was deeply touched by your manifestations of grief."

"Oh! You saw me at the house," exclaimed the bereaved Gaul. "Many thought that fine. But you should have seen me at the grave. There I raised Hell."

Very much in the same way, there was one man at Ralston's obsequies, conspicuous for his ostentatious sorrow, who was more responsible for his downfall than anyone else and profited largely by his death. But after the funeral he was able to speak of the tragic event with much fortitude and a certain degree of complacency. Ralston's death, he said, was extremely opportune—in fact the best thing that could have happened, for it made easy going for everyone.

But nothing can be more true than the cynical words that Shakespeare puts into the mouth of Marc Antony. The evil a man does lives after him. The good is buried with his bones. The city went about its business, forgot its sorrow, which is necessary and proper unless the world is to be draped with perpetual mourning weeds, forgot much of the great services Ralston rendered California, although to this day, among the old-timers and their descendants, his name still stirs a thrill. But all his human weaknesses have been remembered and handed down, duly magnified, to posterity.

Not only that, but his memory has been assailed by accusations of the gravest nature, relating to the failure of the Bank of California. These charges reached me in Kentucky, and as they did not proceed from an authoritative source and, moreover, seemed totally inconsistent with the character of my old friend, I made a special visit to the Pacific Coast to investigate the circumstances immediately preceding and associated with his death for the better satisfaction of myself and of the world at large.

CHAPTER XXXVII

Testimony of Eyewitnesses and Experts Refutes Story that Wm. C. Ralston Took His Own Life

Ruined Financier Had Deeded His Property to William Sharon, Who Forces Widow to Accept $250,000 as Payment in Full

Among the common traditions of William C. Ralston's death is the story that he committed suicide to escape exposure. Notwithstanding the fact that a coroner's jury found on ample expert evidence that he died from a cerebral attack and the further incident that a life insurance company promptly paid a policy of $50,000 to his widow—a policy void by express terms in the event of suicide—this impression seems to persist today.

When I came to California for firsthand information concerning my old friend's tragic end, my earliest business was to investigate the question of self-destruction, for if it were a fact that he made away with himself at a time when much explanation was needed, it would have had assuredly an ugly look. The evidence was all fresh and so overwhelmingly conclusive of death from natural causes that I cannot see on what basis a theory of suicide was reached, unless it were suggested by ulterior motives. The testimony of eyewitnesses was that the swimmer suddenly collapsed and floated with the tide. The lungs were inflated with air, not with water, as in cases of drowning; otherwise, the body would have sunk. The features had not the ghastly pallor that follows water suffocation; on the contrary, they were suffused and livid, as when death ensues from a bursting blood vessel in the brain. To this physicians gave further testimony. The sad facts were plain enough. For many days Mr. Ralston had suffered a mental strain, against which the human machinery is not often proof. It reached the crisis when he was excluded from the meeting of the bank's trustees. Then something snapped. Perhaps the plunge in the cold waters of the bay hurried on the catastrophe. But the baseless story of Mr. Ralston's suicide ought to be finally set at rest.

The inside history of the failure of the Bank of California in 1875 has never been told. About the only definite statement ever made was that its capital stock of $5,000,000 had been exhausted, although the institution had resources sufficient to protect depositors. It was rehabilitated by an assessment of $100 a share, which was paid by the stockholders, giving a new capital of $3,000,000. Five weeks after the failure it reopened its doors, with almost undiminished prestige, and with all the many ups and downs of finance has maintained its position as the leading commercial bank west of the Missouri River.

There is no doubt that Mr. Ralston owed large sums of money to the bank, growing out of many investments, some of which were disastrous. In those days, whatever may be said of the practice, it was the commonest thing for bank officers to make loans to themselves. Not only that, but the practice was in full swing down to the time of the failure of the Safe Deposit Company's bank. With vast visible personal resources as security for loans, Mr. Ralson's unlimited credit never seems to have been questioned by the directors of the Bank of California. Among his assets were a half-interest in the Palace Hotel, a half-interest in the Montgomery Street Land Company, which he and I organized, including the Grand Hotel and most of the frontage on New Montgomery Street, one-half of the capital stock of the Spring Valley Water Works, one-half interest in the Union Milling and Mining Company, which controlled the reduction of ores on the Comstock Lode with enormous profits, and one-third of the stock of the Virginia and Truckee Railroad, which holds the record of earning more per mile than any railroad in the world before or since. I should say that a conservative estimate of these properties alone was not less than $15,-000,000. Besides, he had numberless industrial investments, residences, and immense acreages of real estate in various parts of California.

After I left San Francisco in the early part of 1873, Mr. Ralston engaged in many costly projects. One of these was the purchase of the Catholic Church property on Market Street and the construction of the Palace Hotel thereon by himself and William Sharon. This alone tied up $3,000,-000 of ready money on his account. It is known that he lost heavily on a large purchase of stock in the Ophir Mining Company upon false information that the great Flood and O'Brien bonanza extended into its territory. Several months before the failure he saw that he was deeply involved. Mr. Ralston realized too late that he had gone too far, that he was beyond his depth. He made efforts to secure money on his great holdings. But for evident reason he was compelled to go slow. The spectacle of Ralston as a borrower would have started suspicion at once. Loans were attempted through outside agents, but the door of accommodation was closed. Rumor has attributed this to the manipulation of the Bonanza firm, but a much better reason can be given than that. Nearly all the ready capital on the Pacific Coast was tied up in the wild Comstock speculation, still at its height. There was no money available for legitimate investments or loans of any kind. On top of this came the withdrawal of many large accounts. On the day of the failure more than $1,-

000,000 were unexpectedly checked out. Under this last blow the bank went down. So far as the Bonanza firm was involved, its members were personal friends of Ralston, though not of William Sharon.

Four days before the failure Mr. Ralston made a deed to William Sharon, conveying "all and singular my real and personal property situated in the city and county of San Francisco and the county of San Mateo and elsewhere and wheresoever and howsoever situated, to be managed, sold, and otherwise disposed of for our joint and several interests."

What was the disposition of this vast property? No one will ever know. Was part of it used to repay any indebtedness of Mr. Ralston to the bank? Again the record is silent.

Some years later Mr. Ralston's widow, who is still living, brought a suit against William Sharon for an accounting under the deed of her deceased husband. After some delay, Mr. Sharon filed a general answer to the effect that the property coming from Mr. Ralston into his hands was worth about a million dollars less than nothing. But he offered $250,000 in full settlement, and to avoid the endless delay of litigation and expenses that she did not have the means to meet, this adjustment was accepted by Mrs. Ralston.

Never in the whole history of finance has such a mystery attached as surrounds the failure of the Bank of California and the disappearance of Ralston's fortune. My own belief is that had his life been spared another month, he would have emerged from all his difficulties with a clean sheet. James R. Keene, a trustee of the Bank of California and the only one who seemed inclined to speak, made a printed statement that upon an examination of Mr. Ralston's assets, he was justified in stating that they were sufficient to pay all his debts of every kind and leave a balance of $3,000,000 to his family, and it is worthy of remark that Keene had a very clear business head.

No one seemed anxious to know the facts. Quite the contrary was the case. When I proceeded to gather information of a most important character, J. D. Fry, uncle of Mrs. Ralston, commanded me in the name of the family to cease. To this I had to bow. But Mrs. Ralston is here to say whether her kinsman was her faithful friend or not.

Even Mr. Ralston's private papers and personal accounts were seized, and disappeared. In some ways he was a secretive man. Once he asked me to send a check to A. A. Cohen for $5,000 each month until ordered discontinued. I followed instructions till the total amounted to over $100,000, yet I never had an inkling what the payments were for. Thus it was currently believed that he had many interests in other people's names. Mrs. Ralston well remembers that her husband took her to inspect a fine business block in process of completion, which he told her was his. After his death the title to this same property stood in the name of another, and in that family name it stands today.

There were many wild rumors of wrongdoing that followed the failure of the Bank of California. The only one deserving notice is this: that Mr. Ralston overissued and marketed stock of the Bank of California. Without any evidence to support it that would be received in any court, this unfounded charge has received an astonishing credence, for I can find nothing to support it except absolutely irresponsible hearsay. Besides, it is contradicted by unquestioned facts.

Mr. Ralston had no reason to overissue any stock. He had oceans of prime securities. What he needed was cash, not certificates of shares. His 50,000 shares of Spring Valley alone had a selling-market value at that time of $5,000,000, nearly twice as much as the bank started business with later on. The money simply wasn't in the town to realize on even that splendid security. That's how Ralston and the bank went down.

In a way, my acquaintance with Mr. Ralston was somewhat tragic. With the best intent on either side, something always went wrong. It changed the whole character and purpose of my life. But I only recall him as a most loyal consistent friend, a financier with a very nice sense of honor, and an exemplar of candid courtesy. It seems to me the time has come when tardy justice should be done to the memory of one of California's most illustrious pioneers, who loved his state as no man of station has loved it since, and to whom the present generation owes much.

CHAPTER XXXVIII

Author Tries Luck in Wall Street and Makes Big Fortune, Only to Lose It in Mining Investments

Silver Falls and Land Slides, But Disaster Fails to Discourage Man Who Has Outlived Old Associates

All the various people of the story have been accounted for and decently retired. Before the curtain falls, I have just a word to say about myself to those who have followed the narrative of sixteen tempestuous years from 1857 to 1873.

The role of a Kentucky country gentleman was not to my liking. As I have said, I sold out everything and retired from California after the bursting of the diamond bubble. I resolved that nothing should tempt me again into an active career. But the lure of the busy world was more potent than I realized. After a few years of the simple life I made my headquarters in New York, studied and grasped the investment and speculative markets, and became one of the recognized figures of Wall Street.

Good fortune, as a rule, attended my ventures. Sometimes the tide turned the other way, but I think, taking one year with another, each saw my assets materially increased. At one time I was worth very near, if not quite, four million dollars, which is distinctly more than any man ought to have. But when one is fairly gone in the money-making intoxication he never knows when to stop, any more than the victim of alcohol.

When I was at the zenith of my good luck, I was induced to invest in two mining properties in the United States of Colombia, South America. One was an immense

ASBURY HARPENDING

At the period of his Wall Street operations

silver-mining district, the other a great hydraulic proposition with almost fabulous gold-bearing gravel resources. Both were passed upon favorably by the ablest experts that money could hire. The reports were justified by the facts, yet both projects ended in ruinous disaster for me.

I was drawn into a much larger investment than I contemplated. As I developed the silver property, the economy of a much larger plant and the ownership of adjacent mineral territory became self-evident. There appeared no element of risk. Silver, after various fluctuations, seemed to have reached a firm level. Financial experts were in accord that the price of the white metal could not possibly go lower, was much more likely to advance than to recede. Even with silver at eighty cents an ounce, the profits on my mining operations would be enormous. I figured to clear such profits that in a few years I would receive back my capital investment and own a property with an earning capacity of millions. In fact, during the period of practical operation, these estimates were fully borne out.

Then something happened. Without a note of warning to the commercial world, Great Britain closed the mints of India to the coinage of silver. As long as this vast Oriental market was open, the value of silver was secure. When it closed, like the snapping of a trap, a panic followed which did not end until the price was squarely cut in two. I could not produce an ounce of bullion without an actual loss. An immense investment became instantly valueless. Nearly three million dollars vanished into thin air with the scratching of a pen.

The hydraulic mining project fared no better. The gold-gravel deposit appeared humanly inexhaustible. All the physical conditions seemed favorable. Water had to be brought in a ditch for twenty-three miles. Most of the ditch, carrying 10,000 miner's inches of water, was completed. Then something happened again. For nearly a mile at its upper end, the line of the ditch ran along a rather steep hillside of shale foundation. When the surface was broken, the whole mountain seemed to get in motion. Millions of tons slid down, bringing to naught every effort of our engineers. Money, as a rule, will in the end, conquer every physical obstacle. But about this time a third thing happened, most serious of all: my funds ran so low that to continue the enterprise further meant an invitation to a final and complete disaster.

My fortune was not lost. It is still intact, buried in the mountains of the United States of Colombia. I have no doubt that some adventurous speculator of the future, under happier conditions, will dig it out.

Since then I have been a miner and dealer in mining properties, with the common average of the miner's ups and downs. Much of my time has been devoted to the Mother Lode of California, where I own property that has an immense future.

I am an old man now—in years, but not in hope. I have outlived not alone nearly all my contemporaries covered by this narrative, but the turbulence and ardor of my early years as well. But while many illusions inseparable from the imagination of a robust and enterprising youth have disappeared, I still have very definite ambitions to pull off one more surprise on the world before the close. There may yet be a sequel, another chapter to the story to which may be attached more fittingly than now the sad word that marks the conclusions of all thing human—

(The End)

[*The above was written nearly two years ago. Since then Mr. Harpending's ambition has been realized. He sold one of his mines on the Mother Lode and, after many fluctuations of fortune, is again the possessor of ample means. One of his last and best friends was John A. Finch of Spokane, to whom this volume is dedicated. Just as the forms were going to press, word came of the sudden death of this good gentleman in Idaho. He took great interest in the publication of this book, which he can never read.—Editor.*]

The Western Frontier Library

of which *The Great Diamond Hoax* is Number 10, was started in 1953 by the University of Oklahoma Press. It is designed to introduce today's readers to the exciting events of our frontier past and to some of the memorable writings about them. The following list is complete as of the date of publication of this volume:

1. Prof. Thomas J. Dimsdale. *The Vigilantes of Montana.* With an introduction by E. DeGolyer.
2. A. S. Mercer. *The Banditti of the Plains.* With a foreword by William H. Kittrell.
3. Pat F. Garrett. *The Authentic Life of Billy, the Kid.* With an introduction by J. C. Dykes.
4. Yellow Bird (John Rollin Ridge). *The Life and Adventures of Joaquín Murieta.* With an introduction by Joseph Henry Jackson.
5. Lewis H. Garrard. *Wah-to-yah and the Taos Trail.* With an introduction by A. B. Guthrie, Jr.
6. Charles L. Martin. *A Sketch of Sam Bass, the Bandit.* With an introduction by Ramon F. Adams.
7. Washington Irving. *A Tour on the Prairies.* With an introduction by John Francis McDermott.
8. X. *Beidler: Vigilante.* Edited by Helen Fitzgerald Sanders in collaboration with William H. Bertsche, Jr. With a foreword by A. B. Guthrie, Jr.
9. Nelson Lee. *Three Years Among the Comanches.* With an introduction by Walter Prescott Webb.
10. Asbury Harpending. *The Great Diamond Hoax.* Edited by James H. Wilkins. With a foreword by Glen Dawson.

DATE DUE

RETURNED			
GAYLORD			PRINTED IN U.S.A.